YOUR DATA, THEIR BILLION$

UNRAVELING
AND SIMPLIFYING
BIG TECH

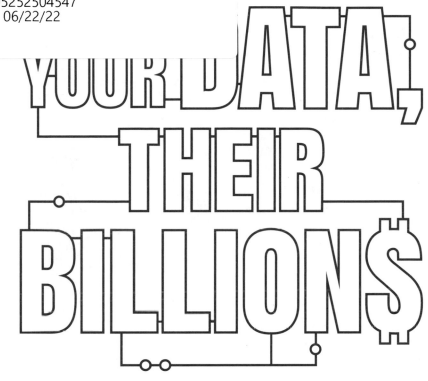

YOUR DATA, THEIR BILLION$

UNRAVELING AND SIMPLIFYING BIG TECH

JANE S. HOFFMAN

Post Hill
PRESS

A POST HILL PRESS BOOK
ISBN: 978-1-63758-074-5
ISBN (eBook): 978-1-63758-075-2

Your Data, Their Billions:
Unraveling and Simplifying Big Tech
© 2022 by Jane S. Hoffman
All Rights Reserved

Cover Design by Cody Corcoran

Post Hill Press
New York • Nashville
posthillpress.com

Published in the United States of America
1 2 3 4 5 6 7 8 9 10

For Michael.

"Science and technology revolutionize our lives,
but memory, tradition and myth frame our response."
—Arthur M. Schlesinger, American historian[1]

---

[1] Arthur M. Schlesinger, "The Challenge of Change", *New York Times*, (July 27, 1986).

Contents

Introduction: The $240 Billion Salad

Lunch. The midday meal. Whether we're grabbing a sandwich from the local deli to eat at our desks, sitting down at a restaurant with a friend or a colleague, or running our kids to the drive-thru on the way to baseball practice, sometime around noon each day it's going to occur to every single one of us to eat something. As a result of the pandemic   living, working, studying, trying to keep ourselves entertained within our own four walls for fifteen months—a good many of us have taken to ordering food in by way of our laptops or an app on our smartphones and having it delivered to our homes or offices. It's convenient, it's quick, and the menu options are nearly unlimited.

The other day, around twelve-fifteen, while I was working at my desk, I decided I wanted my favorite cashew chicken salad from one of my little neighborhood restaurants. So I got out my phone and—as a very old television commercial about the convenience of the Yellow Pages once put it—let my fingers do the walking.[2] What I was working on at my desk was this book—a book about how our personal data is constantly being mined by, and turned into profit for, Big Tech, and what we can do about it. So I hope you'll appreciate the irony when I tell you that the phone knew exactly what I was going to order and prompted me to confirm the delivery address almost as immediately as I'd logged into the app.

I was reminded of a conversation I'd had recently, as I prepared to write this book, with James Waldo, who is the Gordon McKay Professor of the Practice of Computer Science at Harvard. We were talking about

---

[2]   "Yellow Pages 1970," https://www.youtube.com/watch?v=SYpJ1IgGoc0.

Big Tech and the sort of targeted advertising they can do, thanks to the personal data we hand over in exchange for the convenience of shopping online. Jim told me about a men's clothing store, a brick-and-mortar place, now shuttered, in which he used to shop when he was in high school in Salt Lake City, Utah. How a particular clerk at this store always remembered his name and the items he'd bought on his previous trips to the store, and invariably selected items that fit Jim's needs and personal style. He remarked that when a real, live, human being treats us in this way, in-store, we think of it as exemplary customer service, but when a machine makes the same sort of suggestions, appropriate to our needs and our style, we think of it as invasive. "I appreciate the differences," he said to me, "but where do we cross the line on this?" At what point in our retail experience does really good, human, customer service become algorithmic manipulation of our needs and desires? When does a highly personalized shopping experience become intrusive?

That's one of the questions we're going to attempt to answer within these pages. For myself, I found it unnerving that an inanimate object—my *phone*—knew what I wanted for lunch before I could tell it, and, importantly, *it knew where I wanted lunch to be delivered*. Part of our brave, new, algorithmic world is living with the fact that our machines come equipped with GPS—*and they know where we are nearly every second of the day.*

No real, live, human person—not even my husband—knows where I am every second of the day.

But my phone does.

o o o

Now, of course, it isn't my phone that knows so much about me. It's the "Frightful Five"[3]—Facebook, Amazon, Apple, Microsoft, and Google—who have this information. Who know so much more about me than just my affinity for cashew chicken salad.

---

[3]   Farheed Manjo, "Tech's Frightful Five," *New York Times* (May 10, 2017), https://www.nytimes.com/2017/05/10/technology/techs-frightful-five-theyve-got-us.html.

And they almost certainly know at least that much about *you*, too.

Have you ever checked your blood pressure, or your heart rate after a workout, on your watch? Or perhaps opened an email to get your COVID-19 test results? Then you have given the Five all sorts of information about your health. Have you ever watched a movie on your phone while waiting in your doctor's office, or carried a library on an iPad? Then you have provided the Five with valuable clues about your entertainment preferences. Have you ever shopped for groceries online or joined a private alumni group on social media? Then the Five have vital clues about the kind of diet you eat and your level of education. Do you have kids? Have you ever checked your child's test scores online? Or coordinated a play date by way of email? Or, perhaps, in the excitement of finding out you were going to *have* a child, you browsed on your laptop for a crib or other newborn necessities as soon as you got home from the OB-GYN? Then the Five know how many children you have, and what schools you send them to—and that's why your social media feeds and your email inbox start filling up with ads for prenatal vitamins and diaper services before you have even had a chance to tell your family and friends the happy news.

It is from these and other various clues you feed into their algorithms that you, the human, become a set of data points the Five can sell to a company that wants to sell you the latest iteration of their exercise bike or tickets to the newest release from the *Star Wars* franchise, a bottle of the next best miracle diet pill or the hot new baby stroller. This constant exposure to targeted goods and services can foment desires you may never even have known you had. You might have, for example, been perfectly content with the Graco Trax Jogger Click Connect Jogging Stroller you could pick up from Walmart for $139, until you saw the ad for the Chicco Bravo Trio Single Travel System offered by Bed Bath & Beyond for $379.99—the art of upselling, as perfected not by your friendly, local personal shopper, but by your not-so-local algorithm with a cold, dead, electromagnetic heart.

How completely—even creepily—can a cold, soulless algorithm get to know all about you? In order to answer that question, I need you to take a little time-travel trip with me.

We're going back to the year 2012, to visit a Target store. For decades, Target, along with nearly every other large retailer in the world, has tried to collect as much information about its customers as possible. Target does it by assigning, or trying to assign, each customer who walks into a store or shops online a unique "Guest ID number." That helps the store keep track of all the items that customer has purchased, as well as what credit cards they've pay with, if they've used a coupon, or when they've visited the store's website, among other seemingly mundane shopping activities. But how do they use this information they collect? Well, they use it, for example, to send you coupon books in the mail or via email that offer discounts on items they know you've already purchased—say, cleaning supplies when they know you might be running low on a certain product—or that will direct your attention to other areas of their store. The coupons can be highly personalized for each shopper, so, for example, if you regularly buy window cleaner or air freshener at Target, they might send you coupons that will take you to their clothing or home goods or grocery department. The goal here is to expand the habit you already have of shopping for cleaning products at Target, and get you to buy tank tops and oven mitts and milk from them, too, while you're there anyway.

The key word in that last sentence is *habit*. "The reason Target can snoop on our shopping habits is that, over the past two decades, the science of habit formation has become a major field of research in neurology and psychology departments at hundreds of major medical centers and universities, as well as inside extremely well-financed corporate labs. 'It's like an arms race to hire statisticians nowadays,' said Andreas Weigend, the former chief scientist at Amazon.com. 'Mathematicians are suddenly sexy.'"[4] Among the secrets these sexy scientists have found out about our *shopping* habits is that there are certain times in life when habits become flexible—"going through a major life event, like graduating from college

---

[4]   "How Companies Learn Your Secrets," *New York Times* (February 19, 2012), https://www.nytimes.com/2012/02/19/magazine/shopping-habits.html.

or getting a new job or moving to a new town,"[5] or having a baby. If, Target's marketers knew, they could reach a woman early enough in her pregnancy, they could encourage the creation of a new habit for her: shopping in their store. So, they asked Andrew Pole, a statistician who'd been working for the company since 2002: "If we wanted to figure out if a customer is pregnant, even if she didn't want us to know, can you do that?"[6]

Pole could. He did it by moving from merely anticipating a customer's behavior—from figuring out how long a bottle of window cleaner might last in an average household, and sending a coupon to that household when the bottle was down to its last half inch of solvent—to *predicting* it. Pole analyzed the information in Target's massive data trove and "was able to identify about 25 products that, when analyzed together, allowed him to assign each shopper a 'pregnancy prediction' score."[7] Customers who bought, for example, a combination of unscented lotions and soaps, supplements like calcium and magnesium, and extra big bags of cotton balls were likely to be expecting, and that knowledge could then be used to trigger the store's algorithm to send that customer coupons she could use to buy cribs and diapers and baby clothes.

This sort of intensive targeting proved to be "eerily accurate."[8] "An angry man went into a Target outside of Minneapolis, demanding to talk to a manager: 'My daughter got this in the mail!' he said. 'She's still in high school, and you're sending her coupons for baby clothes and cribs? Are you trying to encourage her to get pregnant?' The manager didn't have any idea what the man was talking about. He looked at the mailer. Sure enough, it was addressed to the man's daughter and contained advertisements for maternity clothing, nursery furniture and pictures of smiling infants. The manager apologized and then called a few days

---

5   Ibid.
6   Ibid.
7   "How Target Figured Out a Teen Girl Was Pregnant Before Her Father Did," *Forbes* (February 16, 2012), https://www.forbes.com/sites/kashmirhill/2012/02/16/how-target-figured-out-a-teen-girl-was-pregnant-before-her-father-did.
8   Ibid.

later to apologize again. On the phone, though, the father was somewhat abashed. 'I had a talk with my daughter,' he said. 'It turns out there's been some activities in my house I haven't been completely aware of. She's due in August. I owe you an apology.'"[9]

Among the problems with algorithms taking the place of human, personal shoppers—and, in the process, doing a much more efficiently and eerily *targeted* job of drilling down into your core needs and desires—is just that: algorithms aren't human. They retain the information you feed to them, but they do not put that information into the context of a real, live human life. And they rarely, if ever, update it in accordance with the natural changes that occur in that human life. The data they contain is not distinguished either by real-time changes to a user's real life or by emotional nuance in the way of our interpersonal, and *in-person*, relationships.

Take, as an example, the story of Lauren Goode, a senior writer at *Wired* magazine who cancelled her wedding and, almost two years after the fact, was still being fed wedding ads on Instagram and "a near-daily collage of wedding paraphernalia on Pinterest."[10] She sought out Omar Seyal, head of core product at Pinterest for an explanation as to why social media algorithms seemed so intent on near-daily reminders of a painful period in her life.

"We call this the miscarriage problem," Seyal told her.[11] People who start shopping for wedding venues and attire and cakes tend to actually use the items they're shopping for—that is, the majority go through with a wedding once the planning has begun, he explained. In the same way, people who shop for cribs and diaper genies usually end up with a baby and use those items, too. Seyal explains the issue as "a version of the bias-of-the-majority problem"[12]—people who have a negative experience are part of the minority, and the algorithm doesn't account for this minority experience.

---

9   Ibid.

10 "I Called Off My Wedding. The Internet Will Never Forget," *Wired* (April 6, 2021), https://www.wired.com/story/weddings-social-media-apps-photos-memories-miscarriage-problem/.

11 Ibid.

12 Ibid.

In other words, an algorithm is just an algorithm is just an algorithm. At the end of the day, it isn't going to stop trying to catch a bride's attention with the latest trend in bridal favors, even two years after she's called off her wedding, and it isn't going to stop trying to sell a baby stroller to a couple who has endured a miscarriage.

Algorithms aren't indifferent to emotions, nor are they oblivious to them; they are, by definition, only a set of instructions for solving a problem. A good way to understand algorithms is to think of them as if they were recipes. You open your cookbook and think, *Gee, that recipe for stuffed zucchini looks yummy,* and decide that stuffed zucchini is what you'd like to have for dinner. In order to actually eat stuffed zucchini, however, you have to check your pantry for all the ingredients you need, make a trip to the grocery store to buy the ones you don't, put on an apron and slice some vegetables and bake them in the oven and set the table and so on—the recipe isn't going to make dinner all by itself. The recipe is just a set of instructions; it has no idea what a grocery store or a cutting board or an oven is at all.

An algorithm, similarly, is just a set of instructions (a recipe) you give to a computer, and it uses those instructions to compute against a certain data set (following the recipe in your kitchen) in order to reach some predetermined goal (dinner). Take Spotify as an example. Spotify's goal is to keep the user listening to its playlists, and it does this by cultivating a detailed profile of the user's musical tastes. Its algorithm knows what songs you have liked in the past; it knows what songs are akin to the songs you've liked—perhaps in the same genre, or by the same band, or from the same era; and, importantly, it knows what songs other listeners, who also like the songs you like, have told Spotify they also like, but which you have never listened to before. Wait—what's that again? Let me break it down for you. You have told Spotify that you like the song "Wouldn't It Be Nice" from the Beach Boys seminal album, *Pet Sounds.* The algorithm might take this "knowledge" and recommend a song by Jan & Dean or the Surfaris for your listening pleasure; and it might also notice that a lot of listeners who like "Wouldn't It Be Nice" also like "God Only Knows," also from *Pet Sounds,* but that *you* have never listened to

"God Only Knows" before on its platform. The algorithm, "thinking" it's doing you favor, then slips "God Only Knows" into your rotation. What the algorithm doesn't "know" is that you actively avoid listening to "God Only Knows" ever since your fiancée/fiancé suddenly called off your engagement last spring. It's a song that, on your best day, makes you sad and, on the worst, hits you like a punch in the gut. The algorithm, however, doesn't know this about you, and, furthermore, it doesn't actually care; algorithms don't react to real life and the emotions that real life involves because they don't know what emotions *are*.

Big Tech, however, hasn't let this limitation stop them.

Machine learning is a facet of artificial intelligence (AI). Let's define AI first. AI is a branch of computer science that uses algorithms to simulate human intelligence, including reasoning and decision-making, in machines. Machine learning, then, focuses on creating applications that can improve the machines' function based on the very information they are processing. Algorithms are, as I've just said, merely a set of instructions to follow to solve a problem. In machine learning, these algorithms are "'trained' to find patterns and features in massive amounts of data in order to make decisions and predictions based on new data."[13]

For example, when one of your older photos shows up as a memory on your Facebook feed, it will likely be a photo of a positive event in your life. That's because Facebook's algorithms have been "trained" to recognize and root out certain words, phrases, and images that could remind users of distressing events—words like "miscarriage," phrases like "passed away," and images like one a person might post of the wreckage of her car after it was sideswiped while parked on a busy road. This is not what we think of when we use the phrase "emotional intelligence," but it is certainly emotionally intelligent for a company to go out of its way *not* to remind its customers of troubling times in their lives.

More worrisome—indeed, extremely worrisome to the point of being *chilling*—are technologies that can detect *our* emotions. For an example, let's look again to Spotify. Spotify has recently secured a patent for

---

[13] "Machine Learning," IBM (July 15, 2020), https://www.ibm.com/cloud/learn/machine-learning.

technology that would allow it to identify a user's emotions through voice recognition, then recommend music that fits the user's mood. Let me say up front that while Spotify might hold this patent, the company has so far given no indication that it intends to deploy the technology—but that doesn't stop us from worrying it might, or stewing over the can of worms that could well be opened up if it did. Does the technology work? How accurate is it? And, whether it correctly or incorrectly interprets the emotion you're feeling at any given point in your day, its purpose is to supply you with music that manipulates that feeling. How relaxed are you about having a machine make decisions about your moods?

For that matter, how relaxed are you about a machine making assumptions about your *character* because it "knows" your moods? The *New York Times* "On Tech" writer Shira Ovide asks us to consider what would happen if Alexa or Siri morphed "from digital butlers into diviners that use the sound of our voices to work out intimate details like our moods, desires, and medical conditions. In theory they could one day be used by the police to determine who should be arrested or by banks to say who's worthy of a mortgage."[14] Joseph Turow, a professor at the Annenberg School for Communication at the University of Pennsylvania and author of the book *The Voice Catchers*, has a succinct reply: "Using the human body for discriminating among people is something that we should not do,"[15] though that might well be the path we're now walking.

This could prove to be a dangerous path, indeed. Take, as an example, what Facebook spokesperson Dani Lever described as "an unacceptable error": the company's artificial intelligence software labeling "a group of Black men who appeared in a video shared on the platform as 'primates'".[16] The feature that enabled this unacceptable error was disabled as soon as the problem came to light, of course, but the larger problem

---

14 "Should Alexa Read Our Moods?" https://www.nytimes.com/2021/05/19/technology/alexa.html.

15 Ibid.

16 "Facebook Apologizes for Labeling Video of Black Men As "Primates," *Huffington Post* (September 4, 2021), https://www.huffpost.com/entry/facebook-mislabels-racist-video_n_613345ebe4b04778c005755e.

remains around concerns that the use of AI "can compound racist and sexist stereotypes".[17]

Now, let's not be naïve. Information systems have, historically, reflected the conventions of their times—and those old conventions, unfortunately, are often glaring examples of the evolution of our understanding of human rights and dignities. Take, for example, the Dewey Decimal System, invented in 1873 and still the system under which our public libraries are organized. The system's inventor was a man named Melvil Dewey. "His work led directly to the creation, not just of public libraries in his home state of New York, but to *the entire concept of the free public library in America*. He also invented the Board of Regents in New York, which became a template for public education across the country."[18] Pretty important guy, right? Well, Melvil Dewey was also a notorious racist—so much so that even people in his own day were "appalled"[19] and, in the "1900s, there was an eventually-successful drive to expel him from public life because of his obvious and enormous prejudices".[20] Dewey's prejudices, however, were—and remain—evident in the library cataloging system he invented. As an example, "Each Dewey heading encompasses ten major subjects, dividing each up by subtopics that add digits to the end of the number. Six of the ten subjects in the 200s are explicitly for Christianity-related subjects. Three of those remaining are either explicitly or implicitly Judeo-Christian. Finally, at the bottom of the heap, the 290s cover 'other' religions."[21] Among those "other" religions is the 299.6 subdivision, which covers all "religions originating among Black Africans and people of Black African descent."[22]

---

[17] Ibid.
[18] Anna Gooding-Call, "Racism in the Dewey Decimal System," *Book Riot*, (September 3, 2021), https://bookriot.com/racism-in-the-dewey-decimal-system/.
[19] Ibid.
[20] Ibid.
[21] Ibid.
[22] Ibid.

For over a century, books relating to Black religions and Black religious history were sardined into one-tenth of one percent of Dewey's system and thus were extremely hard for researchers and other readers to access. The internet has made finding works on these subjects so much easier, of course—type "books on Black religious history in your Google search bar and you'll come up, as I did, with over 232,000 results. What lingers from Dewey's day, sadly, is the imposition of an individual's prejudices upon new and emerging systems of organization" and cataloging. This sort of imposition is an important topic, and we'll take it up again later in this book, when we discuss diversity among the staffs of the Big Five. For now, factor into the potential problems that AI poses the age-old one of personal bias among those who actually create the systems we use.

Other potential problems include: the misgendering of transgender people if the technology analyzes voices using male-female binary data;[23] the privacy violations that would almost certainly result from a listening device that is always on and listening for emotional fluctuations in every conversation you have in the privacy of your own home; the security risks associated with third parties, from the aforementioned law enforcement agencies to hackers with unethical intent, who'd like to have access to information about your emotional state—and you easily can see why a recent report, which projects emotion-recognition technologies will be worth $37.1 billion by 2026,[24] is unnerving a lot of people.[25] Technology that has the potential to manipulate us into emotions we might not even be aware we're experiencing has far-reaching consequences.

---

[23] "Computers Are Binary, People Are Not: How AI Systems Undermine LGBTQ Identity," *Access Now* (April 6, 2021), https://www.accessnow.org/how-ai-systems-undermine-lgbtq-identity/.

[24] "The Worldwide Emotion Detection and Recognition Industry Is Expected to Reach $37.1 Billion by 2026," *PR Newswire* (May 13, 2021), https://www.prnewswire.com/news-releases/emotion-detection-and-recognition-market-worth-37-1-billion-by-2026--exclusive-report-by-marketsandmarkets-301256671.html.

[25] KC Ifeanyi, "This Musician Is Calling on Spotify to Ditch Any Plans to Track Listeners' Emotions," Fast Company," (April 9, 2021), https://www.fastcompany.com/90623651/evan-greer-spotify-is-surveillance-emotion-ai.

That said—fully acknowledging the creepier aspects of AI's potential—I do want to be clear that there are any number of potential beneficial uses for the technology. For example, a company called Compology[26] may well be revolutionizing the way industries deal with waste matter, allowing real progress toward sustainability in the area of waste management and recycling. By placing smart cameras paired with AI-powered software directly in dumpsters, Compology allows users to see, measure, and track the waste their own company generates. Redaptive[27] is another company making use of AI, in this case to meter energy consumption and provide real-time data on locations and equipment where excess energy is being siphoned off by inefficient or malfunctioning equipment, decreasing overall energy efficiency and, not unimportantly, increasing your electrical expenses.

○ ○ ○

Returning to the dark side, however, let's also consider what it means for our politics when algorithms feast at the banquet of private data we feed to them.

Have you ever liked the Facebook post of a partisan organization, or retweeted a politician, or even Googled a candidate for office to find out more about her? Have you ever made an online donation to a candidate or a PAC? Signed an online petition? Responded to the solicitation of an interest group, or even a charitable organization, by clicking the link it sent you in an email? Then the Five have a record of just where your politics fall on the spectrum, left to right, and, as a result, the information you find on your social media feed or through your online search is not neutral. Rather, it is specifically curated by algorithms to feed you stories that align with posts you have liked on Facebook and tweets you have retweeted on Twitter and information you have searched for on Google.

---

[26] "Waste Metering and Industrial Dumpster Monitoring," *Compology*, https://compology.com/products/dumpster-monitoring/.

[27] John Schinter, "Six Ways Submetering Improves Energy Insight," *Facility Executive* (February 18, 2020), https://facilityexecutive.com/2020/02/energy-meters-six-ways-submetering-improves-energy-insight/.

The machine decides what you want to hear and read based on what you have *already* heard and read. The machine customizes the content you receive and, because it is a machine, it is agnostic about truth. The information it accesses for you is not necessarily information that has a basis in any reality except the one you have taught the machine to spin for you.

Even more, as a retailer can buy your information from Big Tech in order to sell you movie tickets or diet pills or cashew chicken salad, politicians, political parties, and other interest groups can also buy your information. Based on what you have already told Facebook or Amazon or Google about yourself by the things you have "liked" or bought or researched online, organizations from Greenpeace to the NRA can target your Facebook feed, or recommend a purchase, or curate your search results to reinforce what you are already inclined to believe.

Stripping out from your feed or searches news and information with which you don't already agree results in what internet activist Eli Pariser refers to as a "filter bubble."[28] In a filter bubble you are separated on intellectual, cultural, and ideological levels from viewpoints that differ from those you already hold. This sort of separation doesn't happen only online, of course—in the United States we are reminded nearly every time we watch the news that we're divided into red states and blue states, isolated by our voting patterns, cultures clashing across borderlines[29]—though it is in our digital lives where the divides are reinforced at every turn. Sometimes this sort of isolation is innocuous. Have you liked a song by a hip-hop artist on social media? Then you might be targeted when Saweetie drops her next album but be overlooked when Luke Combs drops his. Sometimes, however, it isn't harmless at all. Have you ever turned to the American Enterprise Institute for information about climate change? Then you're unlikely to find credible scientific news about the climate crisis popping up in your search results. This constant exposure to information—or

---

28 Eli Pariser, "Beware Online 'Filter Bubbles,'" TED.com (2011), https://www.ted.com/talks/eli_pariser_beware_online_filter_bubbles#t-521534.

29 Gus Wezerek, "Do You Live in a Political Bubble?" *New York Times* (April 30, 2021), https://www.nytimes.com/interactive/2021/04/30/opinion/politics/bubble-politics.html.

disinformation, as the case may be—can foment political opinions and goals you may never even have known you had either.

And here's a final kick for you: the more isolated we become in our own, particular political bubble, the more radicalized we become to the beliefs and ideologies of our "side." Setting aside for a moment that radicalization can lead to real danger—see the events of January 6, 2021—let's focus here on the monetization of that radicalization: the more devoted we are to a cause, the more valuable we become to Big Tech, because Big Tech can then turn around and charge the political party or movement with which these radical beliefs are associated even more in ad fees in order to reach you where you live: the internet.

o o o

Bringing this introductory discussion full circle, we come home to the direct link between your data and their dollars: the amount of money the Five rake in from selling your information is staggering. "In the most recently reported fiscal year, Google's revenue amounted to $182.53 billion…. Google's revenue is largely made up by advertising revenue, which amounted to $146.9 billion…in 2020."[30] And that's only Google. "In 2020, Facebook generated close to $84.2 billion…in ad revenues. Advertising accounts for the vast majority of the social network's revenue."[31] That's over $325 billion, generated by only two of the companies in question, in a one-year period. And that's for the year 2020 only. Bloomberg reports, as an example, that in the first quarter of 2021, Facebook's "sales rose 48%, surging past analysts' estimates thanks to strong demand from retailers and other advertisers seeking to grab

---

[30] United States Securities and Exchange Commission, Alphabet Inc. Form 10-K (December 31, 2020), https://abc.xyz/investor/static/pdf/20210203_alphabet_10K.pdf.

[31] "Facebook's Advertising Revenue Worldwide from 2009 to 2020," Statista Research Department (February 5, 2021), https://www.statista.com/statistics/271258/facebooks-advertising-revenue-worldwide.

attention from the social network's billions of users."[32] Not to put too fine a point on it, but that's *your* attention retailers are trying to grab. Those billions that Facebook, Google, Amazon, Microsoft, and Apple are raking in are, in the greatest part, *based on selling your personal information.*

Are you comfortable with that?

o o o

The British historian Ruth Goodman is a specialist in British social history and is currently working on a book about housework—traditionally female activities such as cleaning and cooking and sewing that have kept home fires burning around the world for thousands of years. She recently talked about her project and the necessity of broadening our definition of housework to include those skills that, in the past, women mastered as part of the normal, everyday duties of sustaining a home. For example, in the centuries before off-the-rack clothing was available in every department store, a homemaker needed to know how to sew in order to be able to make her family's clothing. And in the centuries before the existence of mass-produced textiles, she had to know how to grow flax and raise sheep and spin yarn so she would be able to manufacture the very fabrics with which she sewed the clothing her family wore.

Ms. Goodman has also suggested that we will need to redefine housework in terms of those skills that have come to be—or will come to be—necessary to sustain a healthy home into the future. She holds up, as a specific example, the laptops and tablets and smartphones with which the vast majority of modern homes function—and the requirement that a contemporary homemaker needs to be ever more technologically literate in order to maintain the firewalls that keep her or his family's private information both private and secure.

Indeed, Shira Ovide underscores the point in an article bemoaning the on-going spectacle of planned events announcing a tech company's latest, hottest offering: "[technology] is no longer confined to a shiny

---

[32] Naomi Nix, "Facebook's Sales, Users Jump as Pandemic Habits Persist," *Bloomberg* (April 28, 2021), https://www.bloomberg.com/news/articles/2021-04-28/facebook-s-sales-users-jump-as-gains-during-pandemic-persist.

thing in a cardboard box."[33] It is an integral, even workaday, part of our lives, incorporated into nearly every hour. We order the food we eat for breakfast online; we download our children's homework assignments online; we—increasingly—take meetings online; and, when we want to get away from it all, get back out into nature and, say, go for a hike, we decide which trail we want to tackle by first looking up our options online. Moreover, in the last twenty years or so we have witnessed advancements in the capabilities and reliabilities of technology coming at us so fast and furiously that the next innovation, and the next, and the one after that don't seem to be so much revelations as the fulfillments of our expectations.

Our present state of affairs isn't about being awed by technology or figuring out how to incorporate it into our lives, but to decide how to adapt to it in ways that best serve humanity now, and into the future. No less a personage than Tim Cook, CEO of Apple and heir to possibly the grandest and most anticipated planned-reveal events of all time, would seem to agree: "In terms of privacy—I think it is one of the top issues of the century. We've got climate change—that is huge. We've got privacy—that is huge…. And they should be weighted like that and we should put our deep thinking into that and to decide how can we make these things better and how do we leave something for the next generation that is a lot better than the current situation."[34]

I agree completely with Tim Cook. Indeed, my first book, *Green: Your Place in the New Energy Revolution*, was a primer to help people understand the technology behind and economics around clean energy solutions. My goal was to raise awareness around the potentialities and pitfalls of each type of green energy—from solar to geothermal to wind to nuclear and beyond—so that consumers could have a solid,

---

[33] Shira Ovide, "We Don't Need Tech Infomercials," *New York Times (April 14, 2021),* https://www.nytimes.com/2021/04/14/technology/apple-tech-event.html.

[34] Mikey Campbell, "Tim Cook Says Privacy 'One of the Top Issues of the Century,'" *Apple Insider* (January 29, 2021), https://appleinsider.com/articles/21/01/29/tim-cook-says-privacy-one-of-the-top-issues-of-the-century.

well-delineated baseline for participating in the public debate around climate change vis-à-vis our future energy options. In this book, I want to do the same thing for privacy.

Now, the notion of privacy—and how that word will evolve in meaning in our technological future—isn't a subject only Tim Cook and I are interested in, of course. The centrality of the issue to how we will use tech going forward—our confidence in the personal, emotional, political, and financial safety of our online lives—has been on the minds of industry creators and critics alike for many years. Kara Swisher, described by *Newsweek* as Silicon Valley's "most powerful tech journalist," writes frequently about "the surveillance economy that continues to spread like a virus worldwide, even as consumers are less aware than ever of its implications."[35] My goal, once again, is to raise awareness, so those of us who aren't powerful and well-versed tech journalists, or ethicists, or entrepreneurs, but everyday *users* can credibly participate in the increasingly urgent debate around privacy vis-à-vis the technology we grow ever more used to—and dependent upon. Think of this book as your guide to the potential weaknesses in your firewalls, and your family's, and of the actions you can take to shore up those walls so your privacy isn't compromised and those walls are not vulnerable to breach.

○ ○ ○

However, before we get into how your personal information is obtained, categorized, and monetized by the Five—and what you can do to protect your privacy in our Internet Age—let's acknowledge how much we truly do love the internet, and all the really tangible ways it has improved our lives. After all, very few of us are going to respond to the digital privacy problem technology has created for us by simply swearing off tech.

For example, *frictionless* is defined as something that is achieved with no difficulty, as an action taken effortlessly—and it is also a buzzword in the tech world, as they strive to make our online experiences ever more

---

[35] Kara Swisher, "Be Paranoid About Privacy," *New York Times* (December 24, 2019), https://www.nytimes.com/2019/12/24/opinion/location-privacy.html.

effortless and more compelling. They are very good at doing just that. It's too convenient to shop online—who among us hasn't appreciated the ease of one-click, frictionless, online shopping at Amazon, and may soon enjoy the same "frictionless" experience at their in-person stores as well?[36] Who, in all honesty, doesn't think it would be wickedly convenient to drop by the grocery store after a long day at work and pay for her dinner ingredients with a simple swipe of her palm?[37] It's too efficient doing business electronically—can you imagine how the activity of business would slow if we were all suddenly to eschew email and go back to snail mail as the standard? And it's just too pleasurable being able to keep so easily in touch with family and friends. We're not likely to give up the benefits of technology. So, if we can't discard it, we're going to have to fix it.

The first step in fixing the problem of online privacy is consumer awareness. So, let's do a little consciousness-raising. Buckle in; we're about to take a deep dive down the murky rabbit hole of the internet.

---

[36] "Amazon Selling 'Just Walk Out' Frictionless Checkout Platform to Retailers," *Convenience Store News* (March 13, 2020), https://csnews.com/amazon-selling-just-walk-out-frictionless-checkout-platform-retailers.

[37] Jeffrey Dastin, "Amazon to Let Whole Foods Shoppers Pay with a Swipe of Their Palm," *Reuters* (April 21, 2021), https://www.reuters.com/technology/amazon-let-whole-foods-shoppers-pay-with-swipe-their-palm-2021-04-21/.

Chapter 1

What You Mean to Big Tech

The internet is a sort of a miracle. It puts the world, not to mention the entire solar system and beyond, at our fingertips, effortlessly connecting us to information, entertainment, and each other. In other words, in a most basic way, it does exactly what it was meant to do.

Humankind has a long history of innovation and invention in attempts to make communicating with each other easier and more efficient. Think smoke signals, the earliest form of visual communication, first recorded as used in 200 BC to send messages along the Great Wall of China. Think carrier pigeons, first employed in the twelfth century AD to carry information between cities like Damascus, Cairo, and as far away as Baghdad—and more lately pressed into service in World Wars I and II to dependably deliver critical messages to our soldiers across battle lines. Think Pony Express, a failed mail service that operated for just about eighteen months in 1860–1861 in the western United States, but nevertheless played an important conceptual role in the establishment of a permanent, year-round system of transcontinental communication.

Think of the telegraph, which came into general use in the late 1800s and was the world's first form of digital communication, transmitting data from one geographic location to another by way of an electromagnetic medium. For many of us, the idea of a telegraph conjures images from *Downton Abbey*—the Morse code being tapped out with one disembodied finger on the machine's key, the dutiful messenger boy bicycling through the English countryside to deliver the news to the manor house, the silver platter Carson employed to deliver the note to his master. It was an unwieldy process but, at the very least, a huge step up from the Pony Express.

The United States postal service, established in 1775, with Benjamin Franklin as the first postmaster general; semaphore lines; telex machines; mailgrams; big, black, bulky Bakelite telephones with rotary dials and human operators standing by on the other end to connect the call; walkie-talkies and CB radios; answering machines; call waiting and call forwarding; faxes; pagers: all innovations and inventions inspired by our human need to communicate with each other. The contemporary devices that facilitate this communication—smartphones, iPads, and laptop devices with their social media and video-conferencing apps— would indeed seem like a miracle to, say, America's first colonists who depended on transoceanic ships to deliver letters back and forth to loved ones and business associates in Britain, voyages that often took months in either direction. The convenience, speed, and dependability of our current modes of communication would seem miraculous, I dare say, to the average business person as recently as the 1980s, who generally had to wait for a written response to a question or proposal via the daily U.S. "snail mail"[38] post or an interoffice mail system.

Not that the internet wasn't already being conceptualized long before the 1980s. Even a brief history of our digital lives reveals that the fundamentals of *information theory*—the science of quantifying, storing, and communicating information—were mapped out a hundred years ago, in the 1920s, by pioneers like Harry Nyquist and Ralph Hartley at Bell Labs, and focused and unified by the work of Claude Shannon (often called the father of the Digital Age) in his landmark 1948 paper,

---

[38] A slang term most of us have used these days for paper mail delivered via the USPS, alluding to the difference in delivery speed between USPS mail and email. What's worth making a note of is that there was once a proposal for the USPS to deliver your email: https://www.bloomberg.com/features/2016-usps-email/.

"A Mathematical Theory of Communication."[39] The work of these visionaries has been applied in a broad range of subjects, from the exploration of deep space to the invention of the compact disc, from bioinformatics, or the function of molecular codes, to the detection of plagiarism, and, of course, to the development of the internet.

The internet, as we know it now, began to emerge in the late 1950s. At the time, computers were hard-wired directly to terminals with one, exclusive individual user. In 1959, Christopher Strachey—who would go on to become Oxford University's first Professor of Computation—filed a patent for *time-sharing* which, in the computation world, has nothing to do with condos on beaches and multiple "owners," but allows for many users to access a single computing resource at the same time. Strachey discussed his idea with a fellow named J.C.R. Licklider of M.I.T., who proposed a network of such time-shared centers in his 1960 paper, "Man-Computer Symbiosis."[40]

The computers of the 1960s were mammoth and immobile—a user had to either travel to where the computer was physically located in order to work on it, or have magnetic computer tapes delivered to him by way of the snail mail system—but their use was considered essential for national defense. The launch of the Sputnik satellite in October of 1957 led the Department of Defense (DoD) to study ways in which it could continue to gather and disseminate information in the case of a nuclear attack. As a result of these studies, ARPANET (Advanced Research Projects Agency Network) was formed. Originally, ARPANET was accessible only by the DoD and the elite research or academic organizations with whom the DoD maintained contracts. Other, separate networks were created to service early business adapters, like the Prudential Insurance Company, the Nielson Company, and General Electric. It wasn't until 1983, when a new

---

[39] Aftab, Cheung, Kim, Thakkar, Yeddanapudi, "Information Theory: Information Theory and the Digital Age," MIT (Fall 2001), http://web.mit.edu/6.933/www/Fall2001/Shannon2.pdf.

[40] J.C.R. Licklider, "Man-Computer Symbiosis," MIT, (March 1960), https://web.archive.org/web/20051103053540/http://medg.lcs.mit.edu/people/psz/Licklider.html.

communications protocol—the Transfer Control Protocol/Internetwork Protocol (TCP/IP)—was created, that all the existing networks were able to communicate with each other in a universal computer language. Indeed, January 1, 1983, the day the DoD adopted the new TCP/IP standard, is considered the birthday of the internet.

In the less than forty years since the birth of the internet, the marketplace has evolved at a miraculous, if at times alarming, rate. From the UNIVAC (Universal Automatic Computer) of the 1950s and '60s, to today's ubiquitous laptops and tablets, technology has evolved faster than at any other time in human existence—and humans have taken full advantage of its benefits. Students have access to a global supply of resources that a stand-alone campus library could never even hope to have on hand. The innovative use of emerging news and entertainment platforms regularly visited by those same students could foster an appreciation of old-style journalism by whole new generations, as the *Washington Post* is garnering new, young readers with its informative and very funny TikTok account.[41] Electronic medical records, virtual doctor visits, global communication among doctors for the sharing of expertise, patient-doctor messaging, and rapid, often online access to lab results has improved healthcare for wide swaths of populations. Emergent apps could aid in the detection of disease, disorders, and even neurological conditions, such as autism.[42] Self-driving cars and "automated traffic enforcement could make our roads safer and reduce potentially biased police stops of motorists."[43] Our personal time is no longer eaten up by traveling to—and standing in line at—places like the DMV, the post office, the bank, travel agencies, and newsstands because we take care

---

[41] Nicole Gallucci, "Dave Jorgenson Chats About Life as the Washington Post TikTok Guy, His Love of Spam, and More," *Mashable* (March 22, 2021), https://mashable.com/article/dave-jorgenson-washington-post-tik-tok-guy-interview.

[42] "Autism Can Be Detected with New Smartphone App Created by Scientists, *Study Finds* (April 28, 2021), https://www.studyfinds.org/autism-detected-with-smartphone-app-duke-scientists/.

[43] Shira Ovide, "Can Tech Make the Road Safer?" *New York Times* (April 19, 2021), https://www.nytimes.com/2021/04/19/technology/tech-road-safety.html.

of such business online. Our computers have freed us from our desks, while the COVID-19 pandemic has accelerated a nearly global adaption to videoconferencing and, hence, the shift of the workplace from office buildings to our homes.

The pandemic has also accelerated our reliance on the internet at a speed even the most optimistic tech booster could not have seen coming. As early as May 2020, just under two months into the coronavirus crisis, while thousands were being laid off as companies and businesses closed down, Amazon saw so much more demand for online shopping that it hired 175,000 additional employees—and, post-pandemic, is retaining 125,000 of these new-hires as full-time employees.[44] Facebook, for its part, grew from 2.6 billion users in the first quarter of 2020 to 2.850 billion in the first quarter of 2021.[45] At Apple, according to Luca Maestri, the company's chief financial officer, speaking in April 2021, "we have now reached more than 660 million paid subscriptions across the services on our platform. This is up 145 million from just a year ago and twice the number of paid subscriptions we had only two-and-a-half years ago."[46] More customers means more connections and more of the conveniences that only cyberspace can supply.

Big Tech has made plenty of positive changes in our lives, but like a paradox wrapped in an enigma and saved on a corrupted flash drive, it has at the same time wreaked havoc with those same lives and the society, institutions, environments, and economies on which we all depend. How much damage has Big Tech done? We'll get to that. First, we must understand the bottom-line basics that drive Facebook, Amazon, Apple, Microsoft, and Google, which, for the record, is owned by Alphabet Inc.

---

[44] Kate Gibson, " Amazon Says 125,000 of Its Pandemic Hires Can Remain as Ful-Time Workers," *CBS News* (May 28, 2020), https://www.cbsnews.com/news/amazon-pandemic-hires-remain-full-time-workers/.

[45] "FB Earnings Presentation Q1 2021," Investor.fb.com, https://s21.q4cdn.com/399680738/files/doc_financials/2021/FB-Earnings-Presentation-Q1-2021.pdf.

[46] "Apple APPL Q2 2021 Earnings Call Transcript," *Rev* (April 28, 2021), https://www.rev.com/blog/transcripts/apple-aapl-q2-2021-earnings-call-transcript.

These are the Big Five of Big Tech. They're the biggest, richest, and most powerful companies in the world—and their magic formula for success comes down to this very basic equation: your data = their billions.

What Is "Your Data"?

None of the devices that came before our internet-connected laptops, computers, and phones asked us to surrender so much in order to make use of them.

IBM introduced its sleek Selectric typewriter on July 31, 1961, and it was manifestly a game-changer in the communications space. Seven years in development, the Selectric did away with the old-fashioned moveable carriage and the cumbersome basket of individual typebars and gave us a "golf ball" head—known as an "element"—that moved, along with the ribbon, across the page as one typed. The keys themselves required only a tap, rather than a *push*, to get that golf ball moving. The design, created by American designer Eliot Noyes, was positively space age when compared with the machines it replaced. It allowed for speedier, cleaner, and more efficient typing and, by so many contemporary accounts, was a truly revolutionary piece of home and office equipment. But no one—not even the salesperson who sold you your Selectric, if you paid cash for it—needed to know your name, or your Social Security number, or your birthday in order to operate it. Forget about having to have a password to access the project you were using the Selectric to complete. Just roll a fresh sheet of 8.5 x 11-inch paper into the roller, and you were good to go. Passwords were, in those days and in the main, reserved for small boys who wanted to limit access to their treehouses. Even our telephones, hard-wired into our walls, and which theoretically allowed anyone virtual entry into our homes, didn't demand the disclosure of personal information into the general public. Anyone could opt out by choosing to have an unlisted number—albeit "Ma Bell"[47] sometimes required an additional monthly fee for that privilege.

---

[47] "AT&T BREAKUP II: Highlights in the History of a Telecommunications Giant," *Los Angeles Times* (September 21, 1995), https://www.latimes.com/archives/la-xpm-1995-09-21-fi-48462-story.html.

Now, however, as 67 percent of the global population uses smartphones,[48] including 85 percent of Americans,[49] one's information is spread over any number of platforms, and one's individual identity can be uncovered in a whole new variety of ways. It isn't any longer just a matter of being able to be tracked by our names, mailing addresses, Social Security numbers, or phone numbers. Today, content on the internet that can identify us includes our email addresses; social media posts; geolocation, as picked up by the GPS apps on our phones and other devices; IP address, which is the unique internet address assigned to a computer, phone, or other online device; digital images; login IDs; and even our shopping lists. It was the contents of her online shopping cart that identified Lindsey Moers, one of the rioters who caused the chaos at the 2017 "Unite the Right" rally in Charlottesville, Virginia. In reviewing footage of the riot, FBI agents took notice of the weapons in Moers's hands. "After getting her Facebook data, [the FBI] discovered messages from earlier in 2017 referencing unspecified online orders. So they went to Amazon and asked for records associated with her Gmail account. In response to that warrant, Amazon provided [her] previous shopping orders, which contained a baton, pepper spray and a stun gun, bought back in 2015, two years prior to the alleged offenses, according to the FBI's affidavit."[50]

GPS data is also helpful in locating lawbreakers. For example, the Department of Justice (DoJ) used search warrants to obtain GPS and other cell phone records of suspected January 6th insurrectionists from Google. They used this data to investigate Stephanie and Brandon Miller, a married couple who were alleged to have taken part in the Capitol attack. Using the GPS data, combined with information from nearby Wi-Fi access points and Bluetooth beacons, investigators pinpointed the

---

48 "The Mobile Economy 2021," *GSMA*, https://www.gsma.com/mobileeconomy/wp-content/uploads/2021/07/GSMA_MobileEconomy2021_3.pdf.

49 "Mobile Fact Sheet," Pew Research Center (April 7, 2021), https://www.pewresearch.org/internet/fact-sheet/mobile/.

50 "Amazon Gave the FBI the Shopping List of an Anti-Fascist Activist," *Forbes(May 17, 2021),* https://www.forbes.com/sites/thomasbrewster/2021/05/17/amazon-gave-the-fbi-the-shopping-history-of-an-alleged-antifa-activist.

couples' location on that day with an accuracy of within thirty-two feet.[51] GPS is a highly accurate, as well as invasive and controversial, tracking tool: it has been used by federal agencies to enforce immigration laws, by a Catholic news outlet to out a priest who had been suspected of frequenting gay bars, and by the military to locate people who had used Muslim prayer apps.[52] In fact, the location histories of our cell phones—the cell phones of people like you and me—is a stunningly profitable industry in its own right: an estimated $12 billion market.[53]

It isn't only adults, however, about whom we need to be concerned. A Pulitzer Prize-winning investigation for the *Tampa Bay Times* delved deeply into the ways in which the Pasco County, Florida, sheriff's office used data to *guess* who might commit crimes, and then sent deputies to "hunt down" and harass potential criminals.[54] It turned up harrowing details, such as the tracking of children's grades and discipline histories, and how such information was then turned over to the sheriff's office, which used it "to create a secret list of children who could 'fall into a life of crime.'"[55]

Of all these very modern methods of identification, biometric data— such as the voice, fingerprint, and/or facial patterns we use to unlock our apps and devices—is most assuredly the most invasive. These are your actual, physical *body parts* being used to confirm that you are who you say you are. It is already gaining widespread acceptance. In Singapore, for example, it is being tested at airports for use at check-in, immigration

---

51 https://www.businessinsider.com/doj-is-mapping-cell-phone-location-data-from-capitol-rioters-2021-3.

52 Jon Keegan, "There's a Multibillion-Dollar Market for Your Phone's Location Data," *The Markup* (September 30, 2021), https://themarkup.org/privacy/2021/09/30/theres-a-multibillion-dollar-market-for-your-phones-location-data.

53 Ibid.

54 "Targeted," *Tampa Bay Times* (January 2, 2020), https://projects.tampabay.com/projects/2020/investigations/police-pasco-sheriff-targeted/.

55 Romy Ellenbogen, "Pasco School Resource Officers Will No Longer Access Student Data," Tampa Bay Times (May 4, 2021), https://www.tampabay.com/news/pasco/2021/05/04/pasco-school-resource-officers-will-no-longer-access-student-data/.

ports, and as a replacement for human-to-human security checks. Steve Lee, the chief information officer for Changi Airport Group, believes that in the future, facial recognition technology will replace the need for passports. He says, "in the future, you just take your face."[56] And in Australia, facial recognition technology is being used to help identify and locate missing persons, though there is robust debate about the legal and ethical implications. Silkie Carlo, director of digital rights activists Big Brother Watch, says, "It could be the final nail in the coffin for individual privacy and the right to be anonymous in public. Even if some uses are socially well-intended, it is a technology that lends itself to authoritarianism, as we have already seen in the U.K., where it has been used to monitor protests and collect biometric photos of innocent people."[57]

Lest I've led you to believe that this is a problem not impacting us here in the United States, documentation shows that "more than 7,000 individuals from nearly 2,000 public agencies nationwide have used [a company called] Clearview AI to search through millions of Americans' faces, looking for people, including Black Lives Matter protesters, Capitol insurrectionists, petty criminals, and their own friends and family members."[58] And that convenient way of shopping we talked about a few pages ago, where all you have to do is wave your palm to render payment? That's a "biometric authentication system" now in use in several Amazon brick-and-mortar locations, including in the Seattle, Washington, area and midtown Manhattan.[59]

Facial recognition technologies, however, aren't only being used by customs agents and law enforcement agencies on the theoretical hunt for lawbreakers—nor are they being developed for the added "benefit" of

---

[56] "Facial Recognition to Be Used at Singapore Airport," *BBC News Service* (May 1, 2018), https://www.bbc.com/news/technology-43962881.

[57] Ibid.

[58] "Surveillance Nation," *Buzz Feed News* (April 6, 2021), https://www.buzzfeednews.com/article/ryanmac/clearview-ai-local-police-facial-recognition.

[59] Dan Avery, "Amazon Rolls Out 'Pay-by-Palm' in New York City Go Store that Lets Customers Purchase Items with a Simple Wave of a Hand," DailyMall.com (May 10, 2021), https://www.dailymail.co.uk/sciencetech/article-9563651/Amazon-rolls-pay-palm-New-York-City-Store.html.

making shopping easier for us masses. Biometric authentication systems are being incorporated into the surveillance technologies that are being used to monitor those of us who work from home. In April of 2020, at the start of the pandemic, *Washington Post* reporter Drew Harwell asked if the new dream of working from home was actually a nightmare,[60] citing the sort of monitoring software available to employers to track worker productivity, which include keystroke tracking as well as mandating that workers' keep their web cams and microphones on continuously during working hours. In September of 2021, he and colleague Danielle Abril reported on another tool that's been added to an employer's monitoring toolbox: the use of company-mandated facial recognition software. One employee who'd been subjected to this latest tool, attorney Kerrie Krutchik, found that if she looked away from her screen for too many seconds or shifted in her chair, she'd have to rescan her face back in "from three separate angles, a process she ended up doing several times a day."[61] Krutchik described the experience as "just constant, unnecessary, nerve-racking stress: You're trying to concentrate and in the back of your mind you know you're on camera the entire time. While you're reviewing a document, you don't know who is reviewing you."[62] She resigned her job after just two weeks under this sort of constant, intense surveillance by her employer. In the same scenario, what would be your upper time limit on allowing yourself to be spied on so invasively?

I'd be remiss if I left this section in which we talk in such depth about the ways in which technology enables snooping in the most intimate areas of our lives without talking about ATLAS. ATLAS is software developed by the Department of Homeland Security (DHS) U.S. Citizenship and

---

[60] Drew Harwell, "Managers Turn to Surveillance Software, Always-on Webcams to Ensure Employees Are (Really) Working from Home," *The Washington Post* (April 30, 2020), https://www.washingtonpost.com/technology/2020/04/30/work-from-home-surveillance/.

[61] Danielle Abril and Drew Harwell, "Keystroke Tracking, Screenshots, and Recognition: The Boss May Be Watching After the Pandemic Ends," *The Washington Post* (September 24, 2021), https://www.washingtonpost.com/technology/2021/09/24/remote-work-from-home-surveillance/.

[62] Ibid.

Immigration Services (USCIS) to automate the screening of millions of naturalized American citizens. The software allows DHS to run immigrants' case files through federal databases, looking for "indicators that someone is dangerous or dishonest and is ostensibly designed to detect fraud among people who come into contact with the U. S. immigrations system."[63] These indicators include investigating a person's ethnicity and national origin, personal relationships and backgrounds, and biometric information such as fingerprints, and comparing his or her data to that within the FBI's Terrorist Screening Database and the National Crime Information Center, as well as "a variety of unknown sources".[64] In the Trump administration, ATLAS was used as the department's primary background screening system and processed more than 16.5 million screenings and generated about 124,000 cases of potential fraud.[65] "But advocates for immigrants believe that the real purpose of the computer program is to create a pretext to strip people of citizenship."[66] Though a "Privacy Impact Assessment" of the software was compiled and released in October 2020,[67] DHS has been secretive about how ALTAS's algorithm works, as well as the rules the department has in place to determine when an immigrant should have his or her citizenship revoked. President Biden came into office promising more humane immigration policies but, as of this writing, the public is still awaiting both a report from his

---

[63] Sam Biddle and Maryam Saleh, "Little-Known Federal Software Can Trigger Revocation of Citizenship," *The Intercept* (August 25, 2021), https://theintercept.com/2021/08/25/atlas-citizenship-denaturalization-homeland-security/.

[64] Ibid.

[65] "Cuccinelli Announces USCIS' FY 2019 Accomplishments and Efforts to Implement President Trump's Goals," U.S. Citizenship and Immigration Services (October 16, 2019), https://www.uscis.gov/news/news-releases/cuccinelli-announces-uscis-fy-2019-accomplishments-and-efforts-to-implement-president-trumps-goals.

[66] Biddle and Saleh, ibid.

[67] "Privacy Impact Assessment for the ATLAS," U.S. Department of Homeland Security (October 30, 2020), https://www.dhs.gov/sites/default/files/publications/privacy-pia-uscis084-atlas-october2020_0.pdf.

administration that reviews U.S. denaturalization policies, as well as more openness about how ATLAS is used in immigration enforcement.

o o o

Your online content that allows for such ease in identifying exactly who *you* are is technically known as *personally identifiable information* (PII), and it is defined—and protected—in different ways, depending on who's talking about it. To some policymakers, many of the identifiers I've cited here are considered PII and should go hand-in-hand with privacy protections. Director of the Federal Trade Commission's (FTC) Bureau of Consumer Protections, Jessica Rich, explained that the FTC regarded data as "personally identifiable" and therefore deserving of privacy protections "when it can be *reasonably linked* to a particular person, computer, or device. In many cases, persistent identifiers such as device identifiers, MAC addresses, static IP addresses, or cookies meet this test."[68]

Cookies, Not Made by Elves

The *cookies* to which Rich refers have nothing to do with Keebler. They are small bits of information, simple text files, really, that your browser creates and which are stored on your device to keep track of your internet use. They were created to increase the speed of the internet's response to your use, and so make your web experience faster and more pleasant. What cookies are *not* are programs, or viruses, and they can do nothing on or to your computer all by themselves. That last statement, however, comes with one big caveat: some cookies can be used for malevolent purposes. They *can* be used as spyware, and I'll cover the simple but critical ways in which you can protect yourself from these rotten sorts of cookies within this discussion—as well as the repercussions you can expect if you do so. For now, let's focus on the different types of cookies because, like the array of Keebler products on your grocery shelves, not

---

[68] Jessica Rich, "Keeping Up with the Online Advertising Industry," Federal Trade Commission, (April 21, 2016), https://www.ftc.gov/news-events/blogs/business-blog/2016/04/keeping-online-advertising-industry.

every cookie is alike. Some are Fudge Stripes, and some are Sandies, and some are Coconut Dreams.

Let's start with the good cookies. Designed in the 1990s by Marc Andreessen—the inventor of Netscape Navigator, the first web browser for the general public—cookies are what makes internet browsing possible. When you're shopping online, for example, they keep track of what's in your shopping cart; without them, an item would disappear every time you clicked a different item on that website. When you are subscribed to a website, it will also use cookies to remember your login ID and/or your password. Cookies are what allow you to click on your favorites bar and be taken directly to your morning newspaper or the portal of your child's school without having to manually login. They remember articles you've clicked on and what songs, or movies, or documents you've downloaded. Sometimes these benign sorts of cookies are more permanent than at other times. *Session cookies* keep a record of your activity while you're navigating through a website's pages and disappear after you leave that website; *persistent cookies*, also called *tracking cookies*, can remember your activity for years. Session cookies are placed in your computer's active memory, but persistent cookies are set in your hard drive and are not deleted automatically when you end a *session* with a particular website.

I was recently told the story of a man who was an avid grill master. He was idly browsing Etsy for accessories for a grilling method he was thinking of trying—smoking Cornish hens by suspending them whole over the coals on a hanger—but got distracted by a phone call. Before he could purchase any of the items he'd loaded into his cart, he closed the Etsy tab and forgot about the new project. Several weeks later he jumped back on Etsy to see if there were any new barbeque rubs that looked interesting—and tasty. He was delighted to find a brand-new line of rubs based on regional American barbeque flavors. But he was quite surprised when he went to check out and found four different grilling accessories were still waiting there in his cart for him to purchase. He didn't think of it as a *bad* surprise: he was happy to revive the idea of trying out the grilling method that had caught his attention a few weeks prior, and he bought all of the old items along with several jars of the new rubs. And it certainly wasn't bad for the Etsy shops he was patronizing, as several of them made sales they surely would have lost but for the cookie that reminded him he'd once been shopping for a meat hanger. What we do know, however, now that we know a bit more about how cookies are categorized, is that somewhere on his hard drive is a permanent Etsy cookie.

Cookies can also be categorized in terms of their origin. *First-party cookies* are created by the website the user is currently visiting. *Third-party cookies*, however, are set by other websites, different from the one you're currently using, for the purpose of tracking your activity in order to gather data that will allow them to target "relevant" ads to you. The main reason tech giants like Google and Facebook are free to users is because of cookies like this. They make most of their money selling advertising space. The more information they can gather about you, the more highly targeted their ads can become; the more highly targeted their ads can become, the more businesses are willing to pay to advertise on their sites.

As an example, let's say you're an author with a book to sell. If your book were being released in the 1980s, you might take out a newspaper ad or a radio spot and hope that someone who might be interested in your work would see or hear your advertisement before he used the paper to wrap up that evening's trash or changed the station. Today you can take out an ad on Facebook that is served only to people who are *already* interested in your topic. Say your book is a mystery featuring a fifty-five-year-old female private investigator who's cracking a case about a cat thief. You can tell Facebook that the ad is to be served only to women over fifty who like both mysteries and cats. If you were going to release the book only in a digital format, you could add that your ad is to be served only to women who own Kindles. If your book takes place in Vermont, you might also try serving it only to women over fifty who like mysteries and cats and own Kindles and live in New England. The deeper the tech giant can drill down into an individual customer's likes and inclinations, reading habits and geographical location, the more "personalized" an ad can become, and the more businesses are willing to pay for that highly targeted audience.

The ads you see on a Google search results page, Facebook and Instagram news feeds, Twitter, or your Microsoft Edge browser are a direct result of what an algorithm knows about who you are and how you function in the world. These ads are not only creating revenues for tech companies that are larger than some countries, but they are at the heart of the new business model of the digital marketplace.

Seeing customized ads on these free services may seem harmless enough on the surface, but remember that they are being served to you because of *what Big Tech knows about your life*. And the fact that they know so much has ignited a firestorm of controversy about privacy rights. Digital advertising has exploded into a highly sophisticated industry that, thanks to cookies, can follow specific users from website to website.

Take *Flash cookies*, also called *super cookies*, which are independent of any browser. They were designed by Adobe Flash and intended to be permanently stored on your computer—and to remain on your computer even *after you have manually and purposefully deleted them*. These cookies might have been created with the best intentions—to better serve a company's customer base. But the fact is that these types of cookies can interact with multiple websites, collecting your data as easily as you travel from one site to another, and then passing that data along to every website along its way. Some super cookies can contain the name of your computer as well as the file path or directories of key files, and some of them can even override your set cookie preferences and reinstate cookies you have removed.

This latter sort of cookie is called a *zombie cookie*, and for a very good reason. Imagine that you have cleared all the cookies from your computer, but then you visit a website you've also visited in the past. The website is going to check for its cookie and, if it doesn't find it, it will check to see if you've got a Flash cookie. If you do, well, your old cookies, like so many digital zombies, rise from the dead.

Cookies may be anonymous, and many of them seemingly benign, but tech wizards have learned how to combine plain vanilla computer cookies with luscious personal data like usernames, physical addresses, birthdays, and other information we, mostly unwittingly, send out every day into cyberspace to whip up the ultimate, highly valuable delicacy: a specific person's identity.

It's a little unsettling to learn what's really going on below the surface of your web browser, isn't it? One participant in a study about the views of nontechnical internet users, conducted recently by CyLab,[69] Carnegie Mellon University's Security and Privacy Institute, said, "I don't really like the idea of someone looking at what I am looking at, and that kind of freaks me out." Another interviewee added, "It is a little creepy… because I feel that I should get to decide what is going in and out of my computer."[70] Tech security expert and author Bruce Schneier expressed the same sentiment, writing that "ads that follow us around the internet feel creepy."[71] I wouldn't be surprised if most of you reacted the same way when you first realized that—thanks to third-party cookies—the internet remembers what you've been exploring.

Say you were perusing websites for recliners the other night as you watched TV on your thirty-year-old sofa. While you flipped from one furniture site to another and honed in on the models with heat and massage, multiple third-party cookies from every site flowed into your hard drive, ready for the picking by ad tech companies. You, from the comfort of your sagging sofa, have alerted the algorithm that your living room needs an upgrade. Therefore, the next time you log on to the internet, you can expect to have become the target of any number of furniture stores or brands, all vying for the dollars you have set aside to purchase your new sofa.

The upside of being a target and finding all those La-Z-Boy and Ikea and Pottery Barn ads clogging your social media newsfeeds might be that

---

[69] "Cisco Announces Strategic Partnership with Carnegie Mellon CyLab," Carnegie Mellon University Security and Privacy Institute, https://www.cylab.cmu.edu.

[70] Blasé Ur, Pedro G. Leon, Lorrie Faith Cranor, Richard Shay and Yang Wang, "Smart, Useful, Scary, Creepy: Perceptions of Online Behavioral Advertising," https://www.cylab.cmu.edu/_files/pdfs/tech_reports/CMUCyLab12007.pdf.

[71] Bruce Schneier, *Data and Goliath: The Hidden Battles to Collect Your Data and Control Your World.* (New York NY: Norton, 2015), 55.

comparison shopping becomes a bit easier. But even if you're not really bothered that Big Tech knows you need a new sofa, consider that they also know about the times you Googled your old girlfriend, or searched for a local therapist, or looked up how to get rid of a foot fungus, too. A record of your online activity—from the receipt for the iPad you purchased for your niece's graduation gift, to whether or not you've ever visited a porn site—is sitting in a database somewhere in the world. And I do mean it could be *anywhere* in the world: "Using connection tracker LittleSnitch, [Mackeeper] found data from one device was sent to Kansas; Virginia; Dublin; Nairobi; Rio Branco, Brazil; and Taiwan to be stored in databases."[72] Even the most intimate parts of our lives are fodder for Big Tech—and those intimate details may be stored in exotic locations you yourself have never visited.

Consider the case of our cat-mystery novelist. In the course of writing her book, she researched: penalties in New England states for pet-napping; if shipping containers are soundproof; and how to make oleander from one's garden into an efficacious poison. Think of what the internet has decided it knows about her, and you'll appreciate why the meme "I'm-Not-A-Murderer-I'm-A-Writer" is so popular in certain circles.

Former FTC official Jessica Rich described the hidden aspect of all this cookie business as causing "very real" privacy concerns. "Companies are using the data they collect to build highly detailed profiles about individuals, and many of the companies involved in this process are behind the scenes, completely invisible to most of us. Companies also are using high-tech techniques to uniquely identify the devices people carry with them everywhere.... [I]ndustry has started to connect people's digital interactions across the different devices they use in a practice known as 'cross-device tracking.'"[73]

---

[72] Chandra Steele, "Where in the World Are Companies Sending Your Personal Data?" *PC Magazine* (May 5, 2021), https://www.pcmag.com/news/where-in-the-world-are-companies-sending-your-personal-data.

[73] Jessica Rich, "Keeping Up with the Online Advertising Industry," Federal Trade Commission (April 21, 2016), https://www.ftc.gov/news-events/blogs/business-blog/2016/04/keeping-online-advertising-industry.

Cross-device tracking, according to the Federal Trade Commission (FTC), "associates multiple devices with the same consumer and links a consumer's activity across her devices (e.g., smartphones, tablets, personal computers, and other connected devices)."[74] While the FTC acknowledges that this practice helps to make the user's experience more seamless and that it may, in some cases, be useful in fraud prevention, it is often done without the consumer's knowledge or permission—and it does result in the increased collection of personal and sensitive data.

If the fact that your every move is known by hordes of unseen tech marketers is not chilling enough, consider this: tech marketers may not be the only unseen entities with access to your personal information. Cookies create a convenient personalized experience for the average internet user, but they also exponentially increase the vulnerability of your data to attack by hackers. Hackers disguise themselves as legitimate online users, and they tap into the information stored by way of the cookies either to take partial control of a user's online activity, or to steal the user's data outright.

The consequences of cookies—the violations of privacy that can and do lead to exploitation of the garnered private information—have some people applauding the changes that both Apple and Google announced regarding their cookie policies in spring 2021. For Apple's part, the change will require apps that run on its devices to secure a consumer's consent before tracking activity. Google has declared its intent—though without a specific start date—to do away with third-party cookies that use an individual's online behavior to target ads. But let the applause die down and take a look at what these changes will really accomplish.

Apple calls its change App Tracking Transparency, or ATT. The change involves essentially doing away with third-party cookies on its apps—as it has already done within its web browser, Safari—unless the customer buys in and allows the tracking. For Google, it involves

---

[74] "FTC Releases New Report on Cross-Device Tracking," Federal Trade Commission (January 23, 2017), https://www.ftc.gov/news-events/press-releases/2017/01/ftc-releases-new-report-cross-device-tracking.

replacing cookies with what the company is calling Federal Learning of Cohorts, or FLoC, which will group web users within specific interest cohorts. For example, if you regularly visit sites such as Epicurious or Food Network, you'll be designated as a cooking enthusiast and lumped in a group with other foodies. If you regularly check the internet for interesting places to go rock climbing, you'll become part of a group of people that likes travel and extreme sports. Google's idea is that your personal information will be less impacted because you will be hidden among the crowd of people who share your interests. These both might seem like sensible initial steps, but neither is without controversy.

The most prominent opposition to Apple's plan comes from Facebook. It accuses Apple of "hindering digital ad rivals while developing its own marketing business in the background,"[75] and a Bank of America analysis estimates that Facebook stands to lose as much as 3 percent of its revenue because of the change.[76] For Google, "[s]maller rival web browsers, such as Firefox and Opera, have rejected FLoC as an inadequate fix for privacy…. They think FLoC further increases the power of Google, the largest online ad-seller, which has lucrative first-party data from logged-in Gmail accounts and properties like YouTube. Google's ad-tech rivals mostly lack this direct relationship with consumers."[77] Among the other critics of Google's plan are regulators concerned that most people who are being funneled into the FLoC universe aren't aware of their change in status. Owners of larger websites are already developing ways for users to "dodge the system,"[78] and The Electronic Frontier Foundation (EFF), a not-for-profit digital rights organization has created a website, amifloced.

---

[75] Mark Bergen, "Apple and Google Are Killing the (Ad) Cookie: Here's Why" (April 26, 2021), https://www.bloomberg.com/news/articles/2021-04-26/how-apple-google-are-killing-the-advertising-cookie-quicktake.

[76] Ibid.

[77] Ibid.

[78] Matt Burgess, "Google's Plan to Eradicate Cookies Is Crumbling," Wired (April 29. 2021), https://www.wired.co.uk/article/google-floc-trial.

org,[79] which includes a button users can click to find out if they're being covertly tracked by the technology.[80]

While it may look, on the surface, as if these steps are being taken by Apple and Google in an attempt to start to address the problem of online privacy, digging a little more deeply into the details of the proposals reveals that the companies may well stand to gain more than the public will from their implementation.

o o o

Keep in mind, though, that cookies are just one of the ways hackers create doorways into your devices and access your personal information. Let's talk about how common computer hacks really are, and what the scale of the potential damage actually amounts to.

In 2019, a woman named Paige Thompson was arrested for hacking into more than one hundred million Capital One customer accounts. Thompson was a former software engineer at Amazon Web Services, the *cloud service* that was hosting Capital One—a service that delivers computing power through the internet so the customer doesn't have to have the computing resources at her or his physical location. Through Amazon Web Services, Thompson gained access to 140,000 Social Security numbers, one million Canadian Social Insurance numbers, 80,000 bank account numbers, and an undisclosed number of people's names, addresses, credit scores, credit limits, balances, and other information, such as credit card applications that dated as far back as 2005. She gained access by exploiting a misconfigured web application firewall, and, prior to her arrest, attempted to disseminate online the user information she stole.[81] Similarly, a mistake was made in the structuring of

---

[79] "Am I FloCed?" Electronic Frontier Foundation, https://amifloced.org.

[80] Harry Pettit, "AD'S ENOUGH Website Reveals If Google Is SPYING on You with New Hi-Tech 'Ad Tracker,'" *The U.S. Sun* (April 12, 2021), https://www.the-sun.com/lifestyle/tech/2684643/website-reveals-google-spying-ad-tracker/.

[81] Rob McLean, "A Hacker Gained Access to 100 Million Capital One Credit Card Applications and Accounts, *CNN Business* (July 30, 2019), https://www.cnn.com/2019/07/29/business/capital-one-data-breach/index.html.

the Dell system—this one in a component that runs on Dell PCs running Microsoft Windows, and it created a vulnerability that would have allowed a hacker to gain almost total control of the PC. Astonishingly, this mistake sat on the devices for *twelve years* before it was discovered and Dell began efforts to correct it. Happily, if even more shockingly, there have been no reports to date of the weakness being exploited by malicious hackers.[82]

In 2019, a Russian hacking group that called themselves "Evil Corp" took over $100 million from various banks using a type of malware known as "Dridex." Once the malware had infected the systems, it was able to steal login credentials, empty the bank accounts of employees and bank customers, and move the money offshore.[83]

In 2018, the computers of Marriot International were breached, exposing the personal information—credit card numbers, addresses, and passport numbers—of up to 500 million people who had been guests at the company's properties. The hackers, the company conceded, may have had access to such information for up to five years before the hack came to light.[84] In 2013, a hack that CNN Business dubbed "epic and historic" was perpetrated against Yahoo users. Names, email addresses, and passwords were stolen from *three billion* accounts and later found "for

[82] Thomas Brewster, "Warning: 'Hundreds of Millions at Risk' from 12-Year-Old Vulnerabilities Lying Deep in Dell PCs," *Forbes* (May 4, 2021), https://www.forbes.com/sites/thomasbrewster/2021/05/04/warning-hundreds-of-millions-at-risk-from-12-year-old-vulnerabilities-lying-deep-in-dell-pcs/?sh=37a4a3d-63b36&utm_source=newsletter&utm_medium=email&utm_campaign=dailydozen&cdlcid=60853caffe2c195e914a13ad.

[83] Kate Fazzini, "'Evil Corp': Feds Charge Russians in Massive $100 Million Bank Hacking Scheme," *CNBC* (December 5, 2019), https://www.cnbc.com/2019/12/05/russian-malware-hackers-charged-in-massive-100-million-bank-scheme.html.

[84] Nicole Perlroth, Amie Tsang and Adam Satariano, "Marriott Hacking Exposes Data of Up to 500 Million Guests," *New York Times* (November 30, 2018), https://www.nytimes.com/2018/11/30/business/marriott-data-breach.html.

sale on the dark web, a murky network only accessible through certain software."[85]

The web, as we know it and use it every day, is known as the surface web, or the clear web. The "dark web" is actually a part of what is referred to as the "deep web," which consists of sites that won't come up in the findings of a search engine, but are normal, everyday sites nonetheless. The substrata of the deep web that is the dark web, however, is not indexed by search engines, and the sites contained within it are available only with special software. It isn't a place most of us would feel safe browsing, in any case. This is where sites dedicated to animal abuse, narcotics, hitmen, human cannibalism, the sale of illegal firearms, and child pornography flourish. That said, the dark web is also used by activists and whistleblowers who rely on the anonymity of the dark web to carry on their search for truth and justice.

This dark place is also—supposedly—awfully *big*, though there isn't any credible measurement I can cite to tell you exactly how big that might be. Think of trying to assess the acreage of the dark, moonlit, wolf-ridden forest in your favorite fairy tale and you'll have an idea of the difficulty of the calculations. We often think we can access any and every bit of information we may ever need by plugging a question into the Google search bar; but by some estimates, the dark web comprises anywhere from 5 percent, to 48 percent, to fully 90 percent of what is actually available online. Software known as Tor[86]—which is "an acronym for The Onion Router; the onion refers to the layers you go through to disguise your identity"[87]—can be used to anonymize your browsing activity related to normal websites. It does, however, also offer hidden services through

---

[85] Selena Larson, "Every Single Yahoo Account Was Hacked–– 3 Billion in All," *Money* (October 4, 2017), https://money.cnn.com/2017/10/03/technology/business/yahoo-breach-3-billion-accounts/index.html.

[86] Tor Project website, https://www.torproject.org/.

[87] "Going Dark: the Internet Behind the Internet," *National Public Radio* (May 25, 2014), https://www.npr.org/sections/alltechconsidered/2014/05/25/315821415/going-dark-the-internet-behind-the-internet.

which the dark web can be entered and which will, additionally, cloak the user's activity while he or she is visiting there.

Unfortunately, because a user's online activity can be so easily cloaked, cybercrime is only increasing, with FBI Director Christopher Wray likening the crisis of rising incidence of ransomware attacks to the 9/11 terrorist attacks.[88] According to Fareed Zakaria, foreign affairs columnist for the *Washington Post* and CNN contributor, "As we connect more and more things to the internet, all of us become more and more vulnerable to hackers, who can compromise any person or business through the Web and steal their data or freeze them out until they pay a ransom. The pandemic accelerated the transition to a digital economy— and thus accelerated cybercrime. By one estimate, ransomware attacks tripled in [2020]."[89]

Given the acute acceleration of such attacks, coupled with our society's overarching reliance on technology, it isn't at all an overstatement to say that no one—no person and no institution—is immune. In September 2021, for example, just as educational institutions were taking tentative steps toward reopening for in-person learning following the long shutdown necessitated by the pandemic, Howard University, a historically Black research university located in Washington, D.C., suffered a ransomware attack. It was "forced to cancel all of its online and hybrid undergraduate classes" and "the school Wi-Fi network [was rendered] unusable."[90] Colleges and universities are, sadly, particularly vulnerable to cyberattacks. This is because institutions of higher learning must try to balance the need for security against the specific purpose the

---

[88] Aruna Viswanatha and Dustin Volz, " FBI Director Compares Ransomware Challenge to 9/11," *The Wall Street Journal* (June 4, 2021), https://www.wsj.com/articles/fbi-director-compares-ransomware-challenge-to-9-11-11622799003.

[89] Fareed Zakaria, "Cybercrime is putting us on the cusp of a digital pandemic. Here's the way forward," *Washington Post*, (June 10, 2021), https://www.washingtonpost.com/opinions/2021/06/10/cybercrime-is-putting-us-cusp-digital-pandemic-heres-way-forward/.

[90] Josephine Wolff, "Howard University's Devastating Ransomware Attack Can Teach Other Colleges a Valuable Lesson," *Slate* (September 9, 2021), https://slate.com/technology/2021/09/howard-university-ransomware-attack.html.

internet serves in the academic world, which is allowing professors and students as open access as is possible to research and other resources. And they must do this for hundreds, and sometimes thousands of people, all of whom are using different devices in different locations. "'The whole notion of a university is that it thrives on collaboration and exchange of scholarship and ideas, both with people inside the university and outside the university,' [James] Waldo said. 'Building an infrastructure for IT that is based around those assumptions is pretty different from the kind of things that can be done in a corporation where you can dictate to your customer base what they can and can't do, and where you really want to keep the outside out and the inside under control—we can't do either of those things.'"[91]

The ways the unscrupulous can try to use cyber tricks to deceive us aren't limited to ransomware, of course. The story of the foreign prince who emails us to tell us we will be rewarded with millions if only we will wire him a paltry few ten thousand right now is so old and common as to have become a cliché. A few of us have even received emails in the same vein, supposedly from people we know—they were the victim of a hotel theft, all their credit cards are gone, will we please wire a few hundred dollars to the place where they are vacationing so they are able to get home? The elderly parents of a friend were asked to wire money in order to get their grandchild out of jail in a foreign country, and they were so alarmed they complied immediately, before checking to find out their granddaughter was safe in her dorm in a neighboring state.

Those are examples of schemes that go right to your heartstrings. Other schemes are more elaborate and technological. Take the revelation that nearly 2 percent of the thousand highest-grossing apps sold in Apple's App Store are scams, and have bilked customers out of an estimated $48 million—and that, whether the apps are legitimate or not, Apple

---

[91]   Josephine Wolff, "Can Campus Networks Ever Be Secure?" *The Atlantic* (October 11, 2015), https://www.theatlantic.com/technology/archive/2015/10/can-campus-networks-ever-be-secure/409813/.

still takes its 30 percent commission on their sale.[92] Apps with such innocuous names as Guitar Tuner Pro, Mature Dating, and Fun Chart Stickers are among the top eighteen apps that have cheated customers by claiming non-existent affiliations with major brands, such as Amazon and Samsung, and/or have used fake customer reviews to inflate their ranking in the App Store.[93]

Not all hackers are infiltrating our computer systems because they're after money, however. In 2014, in an attack separate from the one referenced in the last paragraph, Yahoo was breached once again, this time by what the company believed was a "state-sponsored actor"—meaning an individual acting on behalf of a government. The personal data—names, addresses, telephone numbers, birthdays, and passwords—of at least 500 million users was compromised.[94] Famously, Russian-backed and Iranian-backed hackers have both tried to influence and disrupt U.S. elections, and arguably succeeded.[95]

And not all private information that suffers compromise is the result of the proactive work of hackers at all. Sometimes it's the result of ineptitude. In 2019, "Two third-party Facebook app developers were found to have stored user data on Amazon's servers in a way that allowed it to be downloaded by the public.... One of the companies stored 146 gigabytes of data containing more than 540 million records, including comments, likes, reactions and account names, on the Amazon servers....

---

[92] Sam Tonkin, "Nearly 2% of the 1,000 Highest Grossing Apps on Apple's App Store Are SCAMS that Have Conned Consumers Out of an Estimated $48 MILLION, Report Warns," *Daily Mail* (June 7, 2021), https://www.dailymail.co.uk/sciencetech/article-9659925/Nearly-2-1-000-highest-grossing-apps-App-Store-SCAMS.html.

[93] Ibid.

[94] Seth Fiegerman, "Yahoo Says 500 Million Accounts Stolen," *CNN Business* (September 23, 2016) https://money.cnn.com/2016/09/22/technology/yahoo-data-breach/?iid=EL.

[95] Jeremy Herb, Brian Fung, Jennifer Hansler, and Zachary Cohen, "Russian Hackers Targeting State and Local Governments Have Stolen Data, US Officials Say" *CNN Politics* (October 23, 2020), https://www.cnn.com/2020/10/22/politics/russian-hackers-election-data/index.html.

Another app is said to have stored unprotected Facebook passwords for 22,000 users."[96]

Other times, it may be the result of lax rules and regulations around the tech space, or the lax enforcement of them. In March of 2018 it was revealed that "Cambridge Analytica, a data firm with ties to Donald Trump, accessed information from as many as 87 million Facebook users without their knowledge."[97] Facebook claimed that the dissemination of this private information was within the rules, as it had been initially collected for academic purposes; only later had it been transferred to Cambridge Analytica and other third parties, in violation of Facebook's policies. One might call this a perfect example of following the letter of the law but totally doing an end run around its spirit.

From manipulating our behaviors and rewiring the concept of who we are, to stolen identities and compromised medical and financial information, to endangering U.S. national security and undermining the fabric of our democracy, the risks of our logged-on lifestyles are many, and ominous.

○ ○ ○

Perhaps one of the most ominous stories—if not *the* most ominous story—concerning how our data is accessed and then put to use concerns an Israeli company called NSO Group, which "develops best-in-class technology to help government agencies detect and prevent terrorism and crime."[98] Now, tracking down terrorists and pedophiles and other criminals sounds like noble work, and it is. But that's not all it is. NSO has a product, commonly known as Pegasus, that it has asserted for years is used by governments only to track criminal activity. But—of course you've guessed by now—governments use the product for purposes beyond tracking criminals.

---

[96] Seth Fiegerman and Donie O'Sullivan, "Hundreds of Millions of Facebook Records Exposed on Amazon Cloud Servers," *CNN Business* (April 3, 2019), https://www.cnn.com/2019/04/03/tech/facebook-records-exposed-amazon/.

[97] Ibid.

[98] NSO Group website, https://www.nsogroup.com/about-us/.

Pegasus is spyware that can be installed remotely on any smartphone. For example, a text message is sent to the phone's owner and, when the owner clicks on a link in the text, the malware is installed. Once installed, it allows the operators in its client countries to take complete control of the targeted phone. This means the operators can turn on the phone's microphone to record its owner's conversations. It can track the phone's location, but it can also turn on the phone's camera to take photos of the owner's location, and people in his or her vicinity. It can access messages, including encrypted ones sent via apps such as WhatsApp. It can acquire the phone's contents, including the owner's contacts, emails, and stored photos.

All this is powerful evidence to use in tracking—in interrupting, and in prosecuting—acts of terrorism and other crimes, to be sure. But in the summer of 2021, a list of around 50,000 phone numbers belonging "to people who are largely based in countries with regimes that are known to spy on their citizens and are also known to be or have been at one time NSO customers"[99] was leaked to Forbidden Stories,[100] a nonprofit journalism organization with its home base in France. The human rights organization Amnesty International then shared the leak with more than eighty journalists from seventeen media groups who worked, under the banner of the Pegasus Project, to identify the owners of those phone numbers.

What they found was staggering. "[T]he list included several heads of state [including French president Emmanuel Macron[101]], cabinet ministers, diplomats, 85 human rights activists, 189 journalists, 65 business executives, military officers and others of note. The latter includes the former wife of assassinated journalist Jamal Khashoggi, and Princess

---

99 Kim Zetter, "The NSO 'Surveillance List': What It Is and Isn't," Zetter.com (July 22), https://zetter.substack.com/p/the-nso-surveillance-list-what-it.

100 Forbidden Stories website, https://forbiddenstories.org.

101 Stephanie Kirchgaessner, "Officials Who Are US Allies Among Targets of NSO Malware, Says WhatsApp Chief," The Guardian website, https://www.theguardian.com/technology/2021/jul/24/officials-who-are-us-allies-among-targets-of-nso-malware-says-whatsapp-chief.

Latifa bint Mohammed al-Maktoum, daughter of Dubai's ruler, who plotted an elaborate escape from her country and family in 2018, only to be captured and returned home."[102]

As of this writing, the leaker hasn't been identified, and the extent of the data breach is yet to be determined. But take a moment to think about what it means when powerful spyware like Pegasus falls into malevolent hands. What it means for the progress of human rights around the world when repressive regimes can purchase malware and use it to spy on journalists and activists and, as in the case of Princess Latifa, their own daughters. What it means for the health of democracies around the world, and for the health of their national security, when heads of state like President Macron—or, perhaps, President Biden—can be remotely spied upon.

What's at Stake

Our daily lives are steeped in a reality that includes, often necessarily and unavoidably, cyberspace. We use an array of devices to digitally organize our time and schedules, our work projects, our children's schooling, and our vacations. These devices are run by data networks—cellular networks, wireless networks, satellite networks, and so on. We willingly—even unthinkingly—take advantage of services offered primarily by Facebook, Amazon, Apple, Microsoft, and Google.

Now, I don't mean to sound judgmental when I use the word "unthinkingly." I mean to say that, at this point in history, cyberspace is woven so seamlessly into our everyday life that *not* using it for the purposes that benefit us would be akin to refusing to use electricity, or giving up indoor plumbing. Think I'm exaggerating? I recently talked with a friend, a doctor, who was planning her family's first post-pandemic vacation, a trip to Disney World. Making arrangements for colleagues to cover her patients; calling in prescriptions so her patients had an appropriate supply of meds while she was away; notifying her children's teachers that they would be absent from their classrooms; helping the kids do the schoolwork

---

[102] Zetter, ibid.

that would keep them current after their return; hiring a house sitter and a dog walker; purchasing new bathing suits for the little ones to wear at the resort pool—and comfortable walking shoes for herself and her husband; booking flights and hotel rooms and an Uber to take them to the airport; choosing from among the almost uncountable restaurants available at the resort, and making the far-in-advance reservations required at the most popular ones: it was all accomplished from her iPad.

Considering how dependent we have become on technology, it's worth taking a deeper look at who *we* are, and what we stand to lose if we pass on the opportunity to check tech while we still have the time.

First, let's do a little breakdown about how we use technology.

- There are over 3.3 billion smartphone users worldwide today. This number is expected to grow by several hundred million in the next few years. China, India, and the United States are the countries with the highest number of smartphone users, with a combined 1.6 billion users.[103]
- In the United States alone, for 2021, the number of smartphone users is estimated to reach 248 million.[104]
- Fully 97 percent of Americans now own a cell phone of some kind, according to the Pew Research Center, and the percentage of Americans who own not only a cell phone but a smartphone, is 85 percent, up from a mere 35 percent in the first survey of smartphone ownership Pew Research Center conducted in 2011.[105]
- The global computer market is projected to grow to $367.56 billion in 2021 from $331.45 billion in 2020.[106]

---

[103] "Insider Intelligence," *eMarketer*, https://forecasts-na1.emarketer.com/5a8601 626ace8d0c50436636/5a4fa637d8690c0c28d1f3a2.

[104] Ibid.

[105] "Mobile Fact Sheet," Pew Research Center (April 7, 2021), https://www.pewresearch.org/internet/fact-sheet/mobile/.

[106] "Computers Global market Report 2021: COVID-19 Impact and Recovery to 2030," *Research and Markets* (January 2021), https://www.researchandmarkets.com/reports/5238063/computers-global-market-report-2021-covid-19.

- In the U. S. alone, 87 percent of the population have access to a computer in their home.[107]
- 4.72 billion people around the world, or close to 60 percent of the global population, make use of the internet.[108]
- The number of internet users is growing throughout the world—332 million new users came online in a one-year period, April 2020 to April 2021.[109] That's 909,589 new users a *day*.
- As of April 2021, it is estimated that 4.33 billion people world-wide, or more than 55 percent of all the people on Earth, use social media.[110]
- The number of people using social media has increased all over the world—by 13.7 percent over the twelve months prior to April 2021 alone. That's a gain of 520 million new users in just one year, or 1.4 million users every single day.[111]
- In 2009, the average American adult spent 2.9 hours a day online. By 2018, that average had gone up to 5.9 hours a day.[112] As of 2021, almost three in ten adults self-report they are online "almost constantly."[113] Indeed, an April 2021 Pew study reported that only 7 percent of Americans don't use the internet.[114]

---

[107] "Insights 2020," Audience Project, https://www.audienceproject.com/wp-content/uploads/audienceproject_study_device_usage_2020.pdf?x54797.

[108] "Digital Around the World," *Data Reportal*, https://datareportal.com/global-digital-overview.

[109] Ibid.

[110] Ibid.

[111] Ibid.

[112] Rob Marvin, "Tech Addiction By the Numbers: How Much Time We Spend Online," *PC Magazine* (June 11, 2018), https://www.pcmag.com/news/tech-addiction-by-the-numbers-how-much-time-we-spend-online.

[113] Andrew Perrin and Sara Atske, "About Three-in-Ten U.S. Adults Say They Are 'Almost Constantly' Online, *Pew Research Center* (March 26, 2021), https://www.pewresearch.org/fact-tank/2021/03/26/about-three-in-ten-u-s-adults-say-they-are-almost-constantly-online/.

[114] Andrew Perrin and Sara Atske, "7% of Americans Don't Use the Internet. Who Are They?" *Pew Research Center* (April 2, 2021), https://www.pewresearch.org/fact-tank/2021/04/02/7-of-americans-dont-use-the-internet-who-are-they/.

- In 2020, Pew reported, 55 percent of American adults either sometimes or often get their news from social media sources. This is up from 2016 when an estimated 42 percent of adults got at least some of their news from Facebook and other social media and compares to the 37 percent who get their news mainly from cable news networks and the 33 percent who get theirs from print.[115]
- In 2020, almost 2.35 billion people worldwide—nearly half of all internet users—purchased goods or services online, and e-retail sales passed $4.2 trillion in U.S. dollars.[116]

An industry that provides services used this extensively by 60 percent of the world's population is an industry with influence. We'll dive more deeply into what Facebook, Amazon, Apple, Microsoft, and Google may be truly worth—and the difficulty of pinning a dollar amount on that figure—when we discuss later in this book what we can know about how the companies calculate taxes owed. For now, an educated estimate is that these five companies have a combined worth of more than $10.7 trillion,[117] giving them a hugely dominant role in the economy and society. Keep in mind, however, that these companies have not only wealth, but global reach—and it is the sort of global reach that is very personal, that reaches right into the pocket or purse where you carry your smartphone every day. That makes these five companies unusually powerful.

---

[115] Tom Infield, "Americans Who Get News Mainly on Social Media Are Less Knowledgeable and Less Engaged," *Pew Research Center* (November 16, 2020), https://www.pewtrusts.org/en/trust/archive/fall-2020/americans-who-get-news-mainly-on-social-media-are-less-knowledgeable-and-less-engaged.

[116] Karin von Abrams, "Global Ecommerce Forecast 2021," Insider Intelligence, eMarketer (July 7, 2021), https://www.emarketer.com/content/global-ecommerce-forecast-2021.

[117] Market cap can change daily; this figure was taken on July 19, 2021.

The dominance of Big Tech makes them "indomitable," in the words of *New York Times* tech writer Farhad Manjoo.[118] These companies, he contends, are more like governments than corporations because they have the wealth to take over huge projects that used to belong to governments—such as spearheading the future of transportation and the development and use of artificial intelligence. They have accumulated all of this wealth and power by creating a feedback loop in which every business or organization that uses computing—which is everybody—must purchase their services.

As Manjoo explains, this loop has only increased their fortunes and clout: "This gets to the core of the Frightful Five's indomitability. They have each built several enormous technologies that are central to just about everything we do with computers. In tech jargon, they own many of the world's most valuable 'platforms'—the basic building blocks on which every other business, even would-be competitors, depend. These platforms are inescapable; you may opt out of one or two of them, but together, they form a gilded mesh blanketing the entire economy."[119]

Not only have these companies created a feedback loop, essentially cornering the tech market, but they also maintain it by either blocking the development of or buying up smaller rivals. Governments around the world have taken, and are taking, stabs at creating new laws and regulations to reign in the power of these companies.

In the United States, during hearings about Big Tech's practices of squeezing out competitors, Senator Amy Klobuchar compared the companies to the abusive powerbroker monopolies of an earlier era, the Gilded Age of the Robber Barons. The Five, she maintained, "have become the kinds of monopolies we last saw in the era of oil barons and

---

[118] Farhad Manjoo, "Tech's 'Frightful 5' Will Dominate Digital Life for Foreseeable Future," New York Times (January 20, 2016), https://www.nytimes.com/2016/01/21/technology/techs-frightful-5-will-dominate-digital-life-for-foreseeable-future.html.

[119] Ibid.

railroad tycoons."[120] By gobbling up start-ups—either by impeding their ability to get a foothold in the marketplace, or through acquisition— Google, Facebook, Apple, Microsoft, and Amazon have, in effect, made it next to impossible for technology entrepreneurs to reach broad audiences.

There is irony in this, of course, as most of these now-enormous companies began with young tech entrepreneurs working in their garages to gain a foothold in the then-nascent tech sector. But as Nicco Mele—the former director of the Shorenstein Center on Media, Politics, and Public Policy at the Harvard Kennedy School and a leading internet strategist—said in his 2013 book, *The End of Big: How the Internet Makes David the New Goliath*, "Radical connectivity is toxic to conventional power structures." Technology has allowed these once-small innovators to be amplified and become powerful, and yesterday's Davids have, indeed, become today's Goliaths. The problem is, as Goliaths, they're keeping their boots on the necks of today's Davids. As the gatekeepers of the technological age, the Five are challenging some of the market principles we value most: they are displaying anti-competitive behavior in ways the oil and railroad tycoons of the past could never have imagined. Move over, Mr. Rockefeller. Take a back seat, Mr. Vanderbilt.

For a final twist in the story of how and why the Five have secured such a potent place in the global economy, let's talk about COVID-19. The pandemic that hit the world in late 2019, and ran rampant for a full year and a half, has, as of this writing, brought with it chaos that has spared no sector—save tech. From restaurants and retailers, to airlines and auto manufacturers, industries almost across the board struggled and shuttered as people in every country on Earth suffered this vicious virus. Schools closed, families were isolated in lockdowns, jobs were lost, anxiety gripped the nations, people died.

But tech?

Tech flourished.

---

[120] Musadio Bidar and Jack Turman, "Klobuchar Pushes for Antitrust Enforcement of Big Tech," *CBS News* (March 19, 2021), https://www.cbsnews.com/news/antitrust-laws-enforcement-big-tech-klobuchar/.

As Peter Eavis and Steve Lohr wrote in the *New York Times* in August, 2020, "A rally in technology stocks elevated the S&P 500 stock index to a record high on Tuesday even as the pandemic crushes the broader economy. The stocks of Apple, Amazon, Alphabet [which owns Google], Microsoft and Facebook, the five largest publicly traded companies in America, rose 37 percent in the first seven months this year, while all the other stocks in the S&P 500 fell a combined 6 percent, according to Credit Suisse."[121] I remind you of the $10.7 trillion combined worth of the five companies we're talking about in this book.

$10,700,000,000,000

Only weeks before the pandemic exploded, the tech sector was under intense scrutiny around several issues, including privacy matters and the dissemination of misinformation, and they seemed vulnerable to the sort of increased oversight they actually require. But the world health crisis shifted everything. If tech had been essential to modern life pre-pandemic, it became an absolute lifeline during the pandemic, and will continue to be essential post-pandemic. We worked from home; our children attended online school. Our social lives were lived over Zoom. We shopped for our nearly every need from e-retailers, and we downloaded apps so we could work out in our living rooms. Our media consumption and social media use skyrocketed—for example, cable news saw up to a 20 percent increase, and daily activity on Twitter went up by 23 percent.[122] While other employers laid off employees or had to let them go entirely, tech remained stable—and was actually hiring. This is particularly true of Amazon, which brought on 175,000 employees in order to meet the volume of online demand.[123] Moreover, experts are telling

---

[121] Peter Eavis and Steve Lohr, "Big Tech's Domination of Business Reaches New Heights," *New York Times* (August 19, 2020), https://www.nytimes.com/2020/08/19/technology/big-tech-business-domination.html.

[122] Media Consumption in the Age of COVID-19," *J.P.Morgan* (May 1, 2020), https://www.jpmorgan.com/insights/research/media-consumption.

[123] Alex Eule and Katie Ferguson, "Big Tech Already Has Great Power. Now, with COVID-19, Come the Responsibilities," *BARRON'S* (May 8, 2020), https://www.barrons.com/articles/big-tech-already-has-great-power-amid-covid-19-it-has-new-responsibilities-51588932007.

us these pandemic-era trends are going to stick as we slowly move into whatever the next "new normal" turns out to be.[124] The pandemic served to solidify tech as not merely a normal part of our lives, but a critical part.

○ ○ ○

Keep in mind, however, that the foundation of tech's success still comes back to what the five tech behemoths know about *you*.

Google knows you well because it scanned your Gmail from 2004 to 2017 in order to better personalize your ad experience. After stopping the practice in 2017, it still allowed hundreds of software companies to do the same. Google has amassed a detailed profile of you based on those emails and your browsing history, the info on your Google calendar, every search you've ever made—and, if you have an Android phone, which Google owns, they know what's in your texts as well. The Android system also shows them who you talk to, what's in your photo albums, where you are right now, and where you've been.

Google, Facebook, and Microsoft know your name, your gender, and your birthday. Facebook knows your physical address, your email address, your phone number, what devices you use, as well as information about your work, income level, race, political views, education, and religion. Google, Twitter, and Microsoft know every video you've ever watched and can identify all your devices. Amazon knows about your devices, too, and it knows your credit card numbers and your government ID.

Google collects personal information about students through its free G Suite for Education that is used by more than thirty million students, teachers, and administrators across the country. In addition to name and date of birth, Google uncovers students' browsing history, location data, contact lists, search terms, and videos they have watched. Half of the school-issued computers students use are cloud-based Google Chromebooks. In a two-year investigation, the Electronic Frontier Foundation discovered that some of the Google programs uploaded

---

[124] "16 Pandemic-Driven Tech Trends that Are Here to Stay," *Forbes* (October 19 2020), https://www.forbes.com/sites/forbestechcouncil/2020/10/19/16-pandemic-driven-tech-trends-that-are-here-to-stay.

"student data to the cloud automatically and by default. All of this often happens without the awareness or consent of students and their families."[125]

The data these companies mine includes more than facts about what we buy and where we go: the real treasure lies in the minute details of our behavior—details that the vast majority of us aren't even aware we're providing.

As Harvard tech scholar Shoshana Zuboff explains, "there is much more behavioral information that we are communicating that we don't know." This includes, of all things, our use of punctuation marks. Exclamation points, bullets, semi-colons, and ellipses—all these marks are used as clues to decipher us and who we are as we innocently type on our devices. With the help of a personality model, these behavioral cues can produce a specific personality profile that marketers can then use to predict how you will respond to messaging. They will then create the ideal message to draw your attention and motivate you to act—buy *this* product, vote for *that* person—based on your personality model.

Zuboff calls this aspect of data mining the "shadow text" because "all of this is happening without our permission, without our knowledge." As she described it to a reporter, the millions of cues that are processed through intricate algorithms construct highly personal stories. "They can tell, for example, if you're gay, or if you're likely to vote alt-right, or if you're likely to be a political malcontent. It's correlated with all kinds of other behavioral predictions. So, for example, they can take all the people who use ellipses, and they can cross correlate that with your outcomes on these personality profiles, and in doing that, they can see that, you know,

---

[125] Gennie Gebhart, "Spying on Students: School-Issued Devices and Student Privacy," *Electronic Frontier Foundation* (April 13, 2017), https://www.eff.org/wp/school-issued-devices-and-student-privacy.

people who tend to use ellipses are people who have uncertainty or don't like to finish things."[126]

The tech industry has learned to "tune, herd, and condition our behavior with subtle and subliminal cues, rewards, and punishments that shunt us toward their most profitable outcomes," Zuboff explained. She refers to Big Tech's model, which has been taken up by many other industries, as *surveillance capitalism*, since it is based on quite literally surveilling us: tracking, claiming, and mining our personal human experience "as free raw material."[127]

All this personal human experience is then processed, sold, and used to predict and shape our behavior for the benefit and profit of someone else. To Big Tech, we are a gold mine. They gather our data for free and sell it to those who not only want to customize their ads to us, but also influence our behavior as we browse online. Are you comfortable with this? Are you okay with being, at the core, not the customer of Big Tech, but its primary product? What are the trade-offs that make the convenience worth it? And, most importantly, are you comfortable with the broader implications of what being a product means?

o o o

Using sensitive personal information to target potential customers with online ads is a hot-button privacy issue, but Big Tech controversies extend into many other areas. The blending of commerce with policing, as an example.

In 2018, Amazon-owned Ring, which makes a popular doorbell camera by the same name, launched a surveillance partnership with police departments across the country. When word leaked about this relationship from a police officer who had attended a Ring training session, civil

---

[126] Noah, Kulwin, "Shoshana Zuboff on Surveillance Capitalism's Threat to Democracy," New York Magazine (February 24, 2019), https://nymag.com/intelligencer/2019/02/shoshana-zuboff-q-and-a-the-age-of-surveillance-capital.html.

[127] Shoshana Zuboff, *Surveillance Capitalism: The Fight for a Human Future at the New Frontier of Power* (New York, NY: Public Affairs, 2019): 8.

rights lawyers began to express their concerns about police officers' ability to look at Ring videos without a warrant.[128] Police need the resident's permission to access the Ring video, but civil rights advocates warned that this mode of warrantless surveillance is a dangerous step. Turning residents into informants who may flag innocent people at their door or across the street as "suspicious," they say, puts citizens at risk.[129] Chris Gilliard, a professor at Macomb Community College in Michigan, cuts clearly to what may be the actual heart of the problem: A Ring camera offers "an opportunity for people to broadcast who they think should and should not be in a particular neighborhood. And often what that means is folks saying that certain Black people, or Black people in general, should not be in their neighborhood."[130]

The product's good intentions of helping neighborhoods fight crime may be overshadowed by these law-and-order concerns over its police relationship, which as of 2020 has extended to 1,400 police departments in cities across the United States, according to the digital rights advocacy organization Fight for the Future.[131] The group's deputy director, Evan Greer, sees the partnership as a reckless marketing move by Amazon that can destabilize our legal processes. "Amazon is building a for-profit surveillance dragnet and partnering with local law enforcement agencies in ways that avoid any form of oversight or accountability that police departments might normally be required to adhere to."[132]

---

[128] Caroline Haskins, "Amazon Told Police It Has Partnered with 200 Law Enforcement Agencies," VICE (July 29, 2019), https://www.vice.com/en/article/j5wyjy/amazon-told-police-it-has-partnered-with-200-law-enforcement-agencies.

[129] Drew Harwell, "Doorbell-camera firm Ring has partnered with 400 police forces, extending surveillance concerns," *Washington Post* (August 28, 2019), https://www.washingtonpost.com/technology/2019/08/28/doorbell-camera-firm-ring-has-partnered-with-police-forces-extending-surveillance-reach/.

[130] Will Oremus, "A Detroit community college professor is fighting Silicon Valley's surveillance machine. People are listening." *Washington Post* (September 17, 2021), https://www.washingtonpost.com/technology/2021/09/16/chris-gilliard-sees-digital-redlining-in-surveillance-tech/.

[131] Cancel Ring website, https://www.cancelring.com/.

[132] Haskins, ibid.

the problem of misinformation that flourishes on social media is as much—or possibly even more—of a problem. As recent elections and the COVID-19 epidemic have shown to us, the proliferation of untruths on the internet is a severe threat to our health, and the health of our democratic system. An informed citizenry is so central to democracy that the guarantee of a free press is enshrined in the First Amendment to the Constitution. Founding Fathers like Thomas Jefferson, who said that "wherever the people are well informed they can be trusted with their own government," would be losing plenty of sleep today over the misinformation that floods the internet.

The well-documented facts of Russian interference in the 2016 presidential election provide evidence that Russian hackers spread inflammatory posts on social media to undermine the Hillary Clinton campaign. The Kremlin-backed hackers also stole files from the Democratic Congressional Campaign Committee's (DCCC) computers and cloud-based service. As one reporter summed up the findings, "the Kremlin mounted a massive online campaign to wreak havoc on U.S. democracy in 2016."[133] The spread of misinformation through media like Facebook and Twitter distracts the public from the facts they need to make informed decisions about *their* government—the government of the people. The ethics of professional journalism demand that the truth be investigated as a check on those we put in power in our local, state, and federal governments, as well as on business leaders. Allowing falsehoods to run rampant online is worse than the cardinal sin of censorship.

As the former executive editor of the *New York Times* and Pulitzer Prize-winning journalist Bill Keller has described it: "Our job—and it's been the job of the press as long as the press has existed—is to help produce an informed electorate so that they can make up their minds. In the case of the war on terror, when you're in one of these really sensitive, fearful times, our job is to let readers know how their government is

---

[133] Eric Geller, "Collusion Aside, Mueller Found Abundant Evidence of Russian Election Plot," *Politico* (April 18, 2019), https://www.politico.com/story/2019/04/18/mueller-report-russian-election-plot-1365568.

doing: Are they doing a good job in protecting the country? You can
that if you allow the government to be your editor or to be your censor.

But neither the government nor any of the scientists who work
for it can protect the country today if false claims about COVID-19
and its treatments spread like wildfire through social media, deterring
mask-wearing or validating vaccine hesitancy. In this case, the misinformation has cost lives as some Americans are convinced that the coronavirus is a "hoax," even denying the existence of the disease on their
deathbed.[135] The late philosopher Dale Jacquette said that "in extreme
cases, the true or false content of a news report can make a difference
between life and death for those who choose to act on what they believe
to be the information the news contains." He emphasized that "it is the
basis for the moral requirement of professional journalistic ethics that
investigators and reporters hold themselves to a high standard of truth
telling."[136] While no news source is perfect—retractions and clarifications
have been offered by even the most venerable of them—old-style media
like national newspapers adhere to those standards; non-professional
organizations and individuals who post "news" on social media to deliberately mislead readers do not.

It isn't only malign foreign actors or hyper-partisan "news" organizations that are responsible for the spread of such damaging misinformation, however; it is also *us*, spreading misinformation to each other,
that is part of the problem. If you have heard one of the crazy rumors
currently making the rounds in any given moment—President Biden is
going to restrict meat consumption in America, or massive quantities of
Vice President Harris's book are being purchased by border patrol officials

---

134 Ken Auletta, "The Press's Role as a Watchdog on Government," *PBS*, https://
www.pbs.org/wgbh/pages/frontline/newswar/tags/watchdog.html.

135 Anagha Srikanth, "'I Thought this Was a Hoax': A COVID-19 Party Guest
Learned Coronavirus Was Real on Their Deathbed," *Changing America* (July 13,
2020), https://thehill.com/changing-america/well-being/prevention-cures/
507046-i-thought-this-was-a-hoax-a-covid-party-guest.

136 Dale Jacquette, *Journalistic Ethics: Moral Responsibility in the Media*. (New York,
NY: Routledge, 2016).

to hand out to refugee children—it is likely you have "heard it relayed from someone you know."[137]

Indeed, I've heard from one friend, who got the information in an email sent to him by a distant acquaintance, that the Democratic-controlled House was going to lift all restrictions on immigration and institute an open-border policy, and, therefore, he was in grave danger of being replaced in his job by an immigrant who would do the work for less pay than he was receiving. "What my acquaintance didn't realize," my friend laughed, "probably because I'm white and don't speak with a discernable accent, is that I pay attention to issues around immigration and that's because my whole family is made up of people who *are* immigrants." I asked if he'd replied to the acquaintance. "Of course," he laughed again. "I wrote him back and asked him if he realized he just sent that filth to a first-generation American. Haven't heard from him since." I laughed with him, but I also wondered how he might have reacted if immigration were not such a personal issue for him, and one in which he was so well-versed. Would he have been able to dismiss the acquaintance's overture so readily in that case? The exchange made me realize how tenuous our hold on truth, as a country, really is.

Facebook and Twitter have upgraded their efforts to remove false information—or, as we'll touch upon in a few paragraphs, have issued public rumblings about efforts to remove false information. Such efforts include tagging or removing false information posts, but their executives are still under fire to explain their failures. On July 28, 2020, for example, Facebook waited several hours to remove a Breitbart News video containing bogus COVID-19 cures and conspiracy theories. The video was rapidly and widely shared tens of millions of times, including by former President Trump on Twitter, so by the time Facebook removed it the video had received more than twenty million views on its platform alone. Facebook stressed that it had been vigilant about removing false content about the coronavirus for months, taking down seven million

---

[137] Max Fisher, "'Belonging Is Stronger Than Facts': The Age of Misinformation," *New York Times* (May 7, 2021), https://www.nytimes.com/2021/05/07/world/asia/misinformation-disinformation-fake-news.html.

false or misleading posts between April and June of 2020, but would need to do a review to uncover "why this took longer than it should have."[138]

In 2018, during one of several hearings that Congress has held with Big Tech, Facebook founder and CEO Mark Zuckerberg admitted that his company should have done more to weed out misinformation and other harmful practices as the company grew. "Facebook is an idealistic and optimistic company," he said. "For most of our existence, we focused on all of the good that connecting people can do. And as Facebook has grown, people everywhere have gotten a powerful new tool for staying connected to the people they love, for making their voices heard, and for building communities and businesses. But it's clear now that we didn't do enough to prevent these tools from being used for harm as well. And that goes for fake news, for foreign interference in elections and hate speech, as well as developers and data privacy. We didn't take a broad enough view of our responsibility and that was a big mistake. And it was my mistake, and I'm sorry."[139]

Being sorry is one thing—and not a bad thing, of course. But raking in billions of dollars while your company hums with misinformation that undermines the fabric of democracy, foments the polarization of a nation's people, and encourages people in taking misguided health risks, is quite another. Putting in place a rigorous policy regarding misinformation would be a more proactive way to offer an apology. Chilling insight into whether Mark Zuckerberg is capable of seeing his way toward a more proactive approach was offered in an October 2021 interview with Kara Swisher: "[Mark] runs a big city that he thinks is running fine, but nobody has water or police or anything else."[140] Even more bluntly, she

[138] Nick Statt, "Facebook Says Removing Viral COVID-10 Misinformation Video'Took Longer than It Should Have,'" *The Verge* (July 28, 2020), https://www.theverge.com/2020/7/28/21345674/facebook-covid-19-misinformation-breitbart-news-video-removal-response.

[139] "How Has Social Media Become So Divisive?" *WNYC Studios* (July 8, 2020), https://www.wnycstudios.org/podcasts/takeaway/segments/how-has-social-media-become-so-divisive.

[140] James D. Walsh, "'The Problem is Him' Kara Swisher on Mark Zuckerberg's crisis and ours.", New York Magazine (October 27, 2021), https://nymag.com/intelligencer/2021/10/kara-swisher-on-mark-zuckerberg-facebook-papers.html.

also observed that "Nobody can do what Mark is doing unless you have an ability to not worry about consequences."[141]

In a story that is unfolding as I type these pages in September of 2021, however, we are discovering that Facebook not only did *not* put into place such rigorous self-regulating policies but actually might have done the exact opposite. Frances Haugen, a former product manager on Facebook's Civic Integrity team during the 2020 election, and, for a short time, after that election as well, is blowing the whistle on the company's internal policies—policies that don't comport with the sort of good-corporate-citizen image the company has sought to portray. "The key problem, Haugen argued, is that Facebook's business is built around driving as much engagement as possible from the social network's billions of users, and data shows that social media users engage more with inflammatory content."[142] As Haugen explained, "Facebook over and over again, chose to optimize for its own interests, like making more money."[143] The upshot of Facebook's choices, Haugen argues, is that safety protocols that were put in place as part of the Civic Integrity program were dropped too quickly after the election and that allowed "Extremists [to] subsequently [weaponize] Facebook to plan the Capitol riot. Facebook posts have repeatedly been cited by federal prosecutors in cases against the Capitol insurrectionists."[144] "When they got rid of Civic Integrity, it was the moment where I was like, 'I don't trust that they're willing to actually invest what needs to be invested to keep Facebook from being dangerous,'" Haugen said.[145]

---

[141] Ibid.

[142] Mary Papenfuss, "Facebook Whistleblower Frances Haugen Reveals Identity in Bombshell '60 Minutes' Interview," *Huffington Post* (October 4, 2021), https://www.huffpost.com/entry/facebook-frances-haugen-60-minutes-whistleblower-jan-6_n_615a46b0e4b008640eb5d6b5.

[143] Ibid.

[144] Ibid.

[145] A.J. McDougall, "Facebook Whistleblower Unmasks Herself, Slams Zuckerberg," *Daily Beast* (October 3, 2021), https://www.thedailybeast.com/frances-haugen-unmasks-herself-as-facebook-whistleblower-on-60-minutes?ref=home.

Not that Haugen is the only, or even the first Facebook employee to become concerned about how Facebook was handling problematic content. "One employee who departed the company in 2020 left a long note charging that promising new tools, backed by strong research, were being constrained by Facebook for 'fears of public and policy stakeholder responses' (translation: concerns about negative reactions from Trump allies and investors)."[146]

And not that the United States is the only country whose government has been staged-managed in the interests of Facebook retaining its earning power. "Today in Vietnam, Facebook is allowing its platform to be abused to divide and isolate people. Troll farms and cyber-army brigades roam the platform, manipulating public opinion and drowning out dissent… Similar things are happening in the Philippines, where Facebook is being used to silence dissent."[147] Why is the social media platform allowing itself to be used for such purposes? "[I]n late 2020, Mark Zuckerberg personally caved to a demand from Vietnam's ruling Communist Party to help silence anti-government critics in the Southeast Asian nation… The Facebook CEO reportedly gave in to the demand following a threat that the company could be knocked offline in Vietnam, where it earns an estimated $1 billion in annual revenue."[148]

On a related note—and touching on a discussion about the transparency of tech that we'll start in earnest in the next chapter—we have to acknowledge Ms. Haugen's courage. Facebook, not unlike other tech

---

[146] Alan Suderman and Joshua Goodman, "Facebook Papers Reveal How It Had Been 'Fueling This Fire' Ahead of The Insurrection," *Huffington Post* (October 22, 2021), https://www.huffpost.com/entry/facebook-papers-capitol-insurrecti on_n_61736795e4b03072d6f6e090.

[147] Mai Khoi, "How Facebook is damaging freedom of expression in Vietnam," *Washington Post* (October2, 2018), https://www.washingtonpost.com/news/ global-opinions/wp/2018/10/02/how-facebook-is-damaging-freedom-of- expression-in-vietnam/.

[148] Elizabeth Dwoskin, Tory Newmyer and Shibani Mahtani, "The case against Mark Zuckerberg: Insiders say Facebook's CEO chose growth over safety," Washington Post, October 25, 2021, https://www.washingtonpost.com/ technology/2021/10/25/mark-zuckerberg-facebook-whistleblower/.

giants, has gone to great lengths to keep its internal operations under wraps. As an example, let's talk about "XCheck". Facebook "has built a system that has exempted high-profile users from some or all of its rules, according to company documents reviewed by *The Wall Street Journal*. The program, known as 'cross check' or 'XCheck,' was initially intended as a quality-control measure for actions taken against high-profile accounts, including celebrities, politicians, and journalists. Today, it shields millions of VIP users from the company's normal enforcement process, the documents show. Some users are 'whitelisted'—rendered immune from enforcement actions—while others are allowed to post rule-violating material pending Facebook employee reviews that often never come."[149] The public, however, seldom knows about such internal goings-on. A large part of the reason that the public views Big Tech through a veil is the measures Big Tech takes to keep employees and former employees from talking about what they witness on the job. Yael Eisenstat, a future of democracy fellow at the Berggruen Institute, former elections integrity head at Facebook, CIA officer, and White House adviser, puts it this way: "Unfortunately, most of the people who know the most about the company's inner workings have been willing to speak to the press only anonymously, for fear of retaliation or breach of non-disparagement agreements that are widely used in the tech world."[150]

Additionally, in September 2021, Facebook added another layer to its strategy of presenting itself to the larger world as a benign force in the social media ecosystem: "Project Amplify".[151] What Project Amplify entails is a shift away from the old method of handling the company's

149 Jeff Horwitz, "Facebook Says Its Rules Apply to All. Company Documents Reveal a Secret Elite That's Exempt," *Wall Street Journal* (September 13, 2021), https://www.wsj.com/articles/facebook-files-xcheck-zuckerberg-elite-rules-11631541353.

150 Yael Eisenstat, "Facebook silences the people who know its operations best," *Washington Post*, (August 3, 2021, https://www.washingtonpost.com/outlook/2021/08/03/facebook-nondisparagement-silicon-valley/.

151 Nick Viser, "Facebook Moves to Tout Image Rather than Apologize for Rogue Content: Report," *Huffington Post* (September 22, 2021), https://www.huffpost.com/entry/facebook-image-project-amplify_n_614ab194e4b0efa77f8698d4.

missteps—which was to apologize for doing things like amplifying Russian attempts in the 2020 election and COVID vaccine misinformation and the content of insurrectionists as they were planning the January 6th attack on the U.S. Capitol—and toward amplifying positive news about Facebook itself. Project Amplify includes "proposals to elevate positive stories about Facebook on users' feeds, the site's most popular section, including posts written by Facebook itself."[152] Kara Alaimo, an associate professor of public relations at Hofstra University, calls Facebook's plan "alarming".[153] "The reason this plan is so disturbing is because Facebook is one of the primary places Americans get their news. According to a just-released Pew survey, 40 percent of adults sometimes or often get their news from social media.[154] If Facebook prioritizes articles that it perceives as favorable to its image, users may, of course, see fewer critical stories since they likely will not be given priority and amplification in the same ways as pieces the company likes."[155]

If most of us live in "filter bubbles"—those closed ecosystems we discussed earlier in this book that serve us news and information based on what an algorithm determines we already know and like—Project Amplify is poised to add a human touch to the task of muffling our information diet. Martha Minow, author of *Saving the News: Why the Constitution Calls for Government Action to Preserve Freedom of Speech*, calls the act of sorting us all into different "digital communities"[156] by

152 Ibid.

153 Kara Alaimo, "Facebook's Alarming Plan for News Feeds," *CNN* (September 22, 2021), https://www.cnn.com/2021/09/22/opinions/facebook-project-amplify-news-alaimo/index.html.

154 Mason Walker and Katerina Eva Matsa, "News Consumption Across Social Media in 2021," Pew Research Center (September 20, 2021), https://www.pewresearch.org/journalism/2021/09/20/news-consumption-across-social-media-in-2021/.

155 Alaimo, ibid, https://www.cnn.com/2021/09/22/opinions/facebook-project-amplify-news-alaimo/index.html.

156 Sophie Hills, "Martha Minow's Warning on the Future of a Free Press," *Provincetown Independent* (August 18, 2021), https://provincetownindependent.org/arts-minds/2021/08/18/martha-minows-warning-on-the-future-of-a-free-press/.

feeding us "different versions of events depending on [our social media] profile 'digital gerrymandering'".[157] When a corporation—one that is a primary news source for 40 percent of American adults—seeks to tamper with its own newsfeed, delivering news that better serves its own interests than the truth does, perhaps we ought to call that censorship.

At the same time as Facebook launched Project Amplify, it "also started cutting back on the availability of data that allowed academics and journalists to study how the platform worked."[158] CrowdTangle is a data analytics tool that is owned by Facebook; it is "a kind of turbocharged search engine that allows users to analyze Facebook trends and measure post performance".[159] But Facebook found this sort of transparency inconvenient as it showed that right-wing posts were "getting much more engagement on their Facebook pages than mainstream news outlets,"[160] and the CrowdTangle team was broken up and some employees reassigned to other divisions.

The question becomes, then, how are we, as tech consumers, going to react to the dismantling of CrowdTangle and the emergence of Project Amplify? Are we going to simply buy Facebook's attempts to polish its own reputation and allow them to continue, as Haugen put it in her October 5, 2021, testimony before Congress, "operating in the dark"?[161] Or are we going to demand light? There is an adage that a self-regulated company is an unregulated company; so far, in our short but intense history as a technologically-oriented society, Big Tech has operated without

---

[157] Ibid.

[158] Ryan Mac and Sheera Frankel, "No More Apologies: Inside Facebook's Push to Defend Its Image," *New York Times* (September 21, 2021), https://www.nytimes.com/2021/09/21/technology/zuckerberg-facebook-project-amplify.html.

[159] Kevin Roose, "Inside Facebook's Data Wars, *New York Times* (July 14, 2021), https://www.nytimes.com/2021/07/14/technology/facebook-data.html.

[160] Ibid.

[161] Olivia Solon and Teaganne Finn, "Facebook Whistleblower Tells Congress Social Network Is 'Accountable' to No One," *NBC* (October 5, 2021), https://www.nbcnews.com/politics/congress/facebook-whistleblower-tell-congress-social-network-accountable-no-one-n1280786.

adequate—and, often, really *any*—government regulation. How would you say that's working out?

Oh, the Humanity

The five most powerful companies in the world are redefining who we believe ourselves to be, one click, one purchase, one GPS-guided drive at a time. We may not be fully aware of it, but the driving force behind Facebook, Amazon, Apple, Microsoft, and Google is not our values, but our *value*—our *dollar* value—and a corporate value system that sees us as goods to be sold more than humans to be served. Those things we consider precious because they are so intimately personal—our heart rate after we climb a set of stairs, or the results of that test our doctor ordered; the lunch plans we make with a colleague, or the text chats we have with our best friend; the music we listen to while we work, or the book we read to our child before bed—are being observed by computers and threaded into packets of data that many businesses and organizations are eager to buy. The fortunes built from these transactions are larger than the wealth of entire countries.

Your spontaneous, individual experience in the here and now—where you're standing, what you're saying and to whom you're saying it, the action you're taking based on your latest thought—these make up your very existence. Your personal information is as private as that book you're reading to your child and the text you just sent to your friend. This is your reality. But in our digital society, that bedtime read and friendly text are transformed into binary numeric bits that carry a monetary value. Your life is reduced to a financial transaction, where your identity is given away in exchange for free internet searches, dynamic driving maps, easy access to your medical records, and online shopping.

Whoever mines the moments of your life as raw material for their business has free access to it in this digital world powered by the "buying and selling of me," as former FCC chairman Tom Wheeler puts it. Psychologist Carl Jung, writing about our "desolate" materialistic outlook in *Modern Man in Search of a Soul*,[162] could not have imagined how much farther down that path we would go, arriving in a digital world where our most subtle, private, and ordinary actions are culled for cash.

As we become more aware of what Big Tech knows about us, and the destructive effects of the Big Tech business model, the more outraged we rightly become. But we can channel our outrage into action. In this book, I'll introduce you to the many ways we can be proactive about our digital privacy, and the various organizations you can join to further the movement for digital privacy and fairness.

For the moment, step back with me to take in the big picture of the digital marketplace, ruled by a handful of too-big-to-fail companies that have written their own rules, so we can more fully understand the power these tech giants wield over our lives. A bird's-eye view gives us a clear picture of the dangers in the sort of unfettered power they possess. Albert Einstein once said that it had become "appallingly obvious that our technology has exceeded our humanity." As we sit precariously at our current technological tipping point, it becomes ever clearer that we must put some fetters on the Big Five if we're going to be able to enjoy the myriad benefits of technology, and use them to improve our lives, without sacrificing our privacy, our autonomy, and, ultimately, our humanity.

Big Tech has remade our world in ways both good and bad. The good are myriad. They have made many of the tasks of our lives more convenient, faster, and more efficient, and enhanced the ways in which we entertain ourselves and form community. The bad must be known, addressed, and transformed.

---

[162] Carl G. Jung, *Modern Man in Search of a Soul* (Harcourt, Brace, & World, 1933).

Chapter 2

The Business of Illusion

When we go to see a show at a theatre, we expect illusion. We see a pair of what look like solid gold candlesticks atop an onstage mantle, and we know intuitively they're not really solid gold, only some clever crafting by a prop master—likely the result of a pair of wooden, thrift store candleholders and a can of metallic spray paint. But for the sake of enjoying the drawing room comedy, we suspend disbelief. We hear a shot fired in the distance and, even as the actors react with surprise and concern, we know no actual gun was fired, that the noise was the work of a stagehand off in the wings—and, still, we let ourselves get caught up in the playwright's mystery. We go to a magic show and we *ohh* and *ahh*, even as we fully understand human beings do not float in space without the help of complicated, artfully hidden wires and pulleys. But the deception is breathtaking, and we relish it. We appreciate that these sorts of illusions are risk-free tricks. For the duration of the show, we allow ourselves to buy into the razzle-dazzle.

I've long been familiar with this last sort of illusion—my brother was an amateur magician who performed at kids' birthday parties when he was a teenager. But the razzle-dazzle of Big Tech is another thing entirely. It isn't quite so transparent. Or so harmless.

o o o

In 2019, Belgium's VRT NWS broke the story that Google Assistant was, at times, recording users' voices on their Google Home device or phone *without being prompted.*[163] Now, Google made it clear in their terms and conditions that what a user says to Google Assistant is recorded and stored for analysis, to improve its voice-recognition technology. Nowhere, however, did it state that Google could hit "record" without the customer's explicit command to turn on the device—or that Google employees, rather than Google *computers*, could listen to those recordings to do the analysis. When VRT NWS reporters listened to more than a thousand excerpts recorded via Google Assistant, they could "clearly hear addresses and other sensitive information."[164] That made it easy for them to track down the people who had been recorded and reveal the audio to them. "This is undeniably my own voice," said one surprised man. A couple from Waasmunster, in Belgium, immediately recognized the voice of their son and their grandchild. These were "bedroom conversations, conversations between parents and their children, but also blazing rows and professional phone calls containing lots of private information."[165] Google recorded, and stored, all of them.

Pause here with me for a moment: think about how you'd feel if you heard one of your private conversations was recorded for posterity and stored in the digital files of a tech behemoth. Were you chatting with your brother about what to buy your mom for Mother's Day, or were you talking to him about the imminent need to place your mother in a long-term care facility? Google could have either or both conversations in their records. Were you praising your daughter for the good grades on her report card, or were you in the middle of an intervention regarding an opioid addiction? Google might have that conversation on file, too.

---

[163] Tim Verheyden, Denny Baert, Lente Van Hee, Ruben Van Den Heuvel, "Google Employees Are Eavesdropping, Even in Your Living Room, VRT NWS Has Discovered," *VRT* (July 10, 2019), https://www.vrt.be/vrtnws/en/2019/07/10/google-employees-are-eavesdropping-even-in-flemish-living-rooms/.

[164] Ibid.

[165] Ibid.

Were you discussing intimate family financial matters with your spouse, or perhaps in the middle of something a lot more intimate when your Google device decided to turn itself on? There's no way to know—and there's no way to sugarcoat what the company was doing: it didn't disappoint the public's reasonable expectations of privacy; it eavesdropped on its customers.

Or, how about this? The case of the Oregon woman who took her iPhone into Apple for repair, and found out from a friend that nude photos of herself had been uploaded to her personal Facebook page. The photos had been stored on her phone, and repair techs had taken it upon themselves to make ten of the photos public.[166] "This isn't the first time an Apple employee or contractor has sorted through a customer's photo gallery and shared their nudes," reports *Vice*. "It happens all the time. In 2019 an Apple genius texted himself a personal photograph of a customer who'd come in looking for help with her phone. In 2016, Apple fired a group of employees in Australia after uncovering evidence the group had set up a website to share customer's photographs."[167]

This proof of eavesdropping and invasion were huge blows to the democratic expectations of fairness, honesty, and transparency among the companies' customers. I've always been a stickler for transparency.

When I was a kid, one of my heroes was Betty Furness. In 1967, President Lyndon B. Johnson had asked Ms. Furness to serve as Special Assistant for Consumer Affairs. Her portfolio included working to pass a law to "give installment buyers a clear statement of interest charges, protection for investors in underdeveloped land, and making natural gas pipelines 'as safe as they can be.'"[168] She subsequently served as a

---

[166] Matthew Gault, "She Sent Her iPhone to Apple. Repair Techs Uploaded Her Nudes to Facebook," *VICE* (June 8, 2021), https://www.vice.com/en/article/pkbkey/she-sent-her-iphone-to-apple-repair-techs-uploaded-her-nudes-to-facebook.

[167] Ibid.

[168] "Betty Furness Takes Over as Consumer's Advocate in White House," *New York Times* (May 2, 1967), https://timesmachine.nytimes.com/timesmachine/1967/05/02/83589060.html?pageNumber=28.

board member for Consumers Union, which is the publisher of *Consumer Reports*, and was appointed by then-Governor of New York, Nelson Rockefeller, to be the first chairman and executive director of the New York State Consumer Protection Board. Her television show, *Buyline: Betty Furness*, established Ms. Furness as "one of a handful of first-rate watchdogs in television consumerism,"[169] and the show won a Peabody Award in 1977. In 1973, Ms. Furness became head of the New York City Department of Consumer Affairs.

I was fascinated by her work. When one of her segments was scheduled to air, my family knew they could find me planted in front of the TV, waiting to hear what she had to say. I knew, and from a young age, that I wanted to do what she did. I wanted to become a consumer advocate like her, demystifying and exposing corporate policies and practices so people could be more informed and astute consumers, and helping to change government policies to benefit end users. That goal is what inspired my previous book, *Green: Your Place in the New Energy Revolution*, and it's what's inspiring this one.

Twenty-two years ago, I stepped into the role of commissioner of New York City's Consumer Affairs. One of my proposals in those years was a new policy I called the Cold Snap Pass, which would give low-income New Yorkers a break from paying tax on their heating oil when temperatures dropped below freezing. The genesis of this proposal was the number of calls that came into my office from people who wanted to know why their heating oil cost so much. A guy would show up with a tanker truck to someone's house on Staten Island and tell whoever was home that they needed two hundred gallons of oil. Then the guy would fill the tank and hand the consumer a bill. How could the homeowner be sure he needed two hundred gallons in his tank? And why did he get charged $3.45 a gallon when his brother in Queens just got charged $3.25?[170]

---

[169] John J. O'Connor, "TV: 'Buyline' With Betty Furness," *New York Times* (May 15, 1978), https://timesmachine.nytimes.com/timesmachine/1978/05/15/110953396.html?pageNumber=61.

[170] Numbers have been contemporized.

I started investigating and learned that oil trading on Wall Street had something to do with it, but so did the fact that some oil companies were overcharging. I put together a task force and held hearings to get information from experts and testimony from consumers to arrive at the best solutions for the problem. We reported for the public, spelling out how oil pricing worked and offering tips on how to conserve, and we proposed the Cold Snap Pass to provide a more tangible form of relief to those who needed it most.

The Cold Snap Pass, however, didn't pass. We didn't reboot Wall Street or take down big oil. But what we did was let consumers know what was going on so they could take more control of their oil deliveries, consumption, and, importantly, the bottom line on their heating oil bills. Transparency = knowledge = power.

Old Model, New Reach

In the interest of transparency, let's take a look at one of the oldest illusions in modern advertising, because two of the Big Five—Google and Facebook—have in the main amassed their fortunes in a style as old as Ben Franklin's scheme for selling more newspapers in the early 1730s.

In 1729, Franklin and his partner, Hugh Meredith, purchased the newspaper the *Pennsylvania Gazette*. Under its previous editor, along with providing news about public events, the *Gazette* had been serializing Ephraim Chambers's *Cyclopaedia*, one of the first general encyclopedias to be produced in the English language. The *Cyclopaedia* was dropped under Franklin's management; he saw the newspaper as more of an outlet for essays and letters from readers, as well as a forum in which he could write about his experiments with electricity and his political opinions.

He also saw it as a vehicle with which to make money. He improved the news reporting and used better types, and readership of the paper soared. When he was appointed Postmaster General, he saw the opportunity to increase circulation by using the newly established post office to expand the paper's delivery area, and it became the most popular paper in the colony. He also created a scheme to increase the paper's revenues by keeping the price of subscriptions low and subsidizing the costs of printing and delivery with advertisements, which, in his plan, would become the main source of revenue. Businesses won by reaching more potential customers, and readers won by gaining cheap access to the news of the day.

Franklin's formula has been a winning one ever since: what began with print media evolved into radio programming in the 1920s and television in the 1940s. Network TV ran soap operas like *The Secret Storm*, which the era's traditional stay-at-home wives and mothers watched while doing the ironing, as vehicles for selling household products to women. Today, a pay channel like HGTV offers real-estate programming that creates a dream audience for mortgage lenders and big-box home improvement stores. ESPN offers sellers America's largest audience of men in the 18-to-34-year-old demographic.

The business formula of American media has been remarkably simple and consistent over the centuries, and is sometimes even transparent about its purpose. In the late 1800s, the publisher of the enormously popular magazine *Lady's Home Journal* admitted:

> "Do you know why we publish the *Lady's Home Journal*? The editor thinks it is for the benefit of American women. This is an illusion, but a very proper one for him to have. But I will tell you the real reason, the publisher's reason, is to give you people who manufacture things that American women want and buy a chance to tell them about your product."[171]

Today, Google and Facebook run the same formula. The engine that drives their nearly $231 billion combined worth is advertising revenue. As did Franklin with the *Pennsylvania Gazette*, our new, digital media keeps its services not merely affordable but free to the consumer, while giving

---

[171] John McDonald Hood, *Selling the Dream: Why Advertising is Good Business,* (Praeger, 2005), p. 30.

those people who manufacture things that women and men the world over want to buy a chance to tell them about their products.

Unfortunately, the folks behind Big Tech have not been as forth-coming with us about their true purpose as that long-ago publisher at *Lady's Home Journal*. For example, Facebook CEO Mark Zuckerberg, whose company derives 98.5 percent of its revenue from advertising,[172] recently updated his ever-shifting definition of Facebook's purpose as a "a privacy-focused communications platform" and "the digital equivalent of the living room."[173]

That's a lovely idea, actually: a digital living room where we can meet even our farthest-flung friends and family whenever we want. But in my living room, a conversation with loved ones has never been interrupted every third sentence with an advertisement. Even door-to-door salesmen have gone the way of landlines and VCRs. At the same time, Facebook is expanding its "living rooms" to include new ad-carrying features such as Neighborhoods.

Rana Foroohar, global business columnist for the *Financial Times*, sums it up neatly: "Filter bubbles, fake news, data breaches, and fraud are all at the center of the most malignant—and profitable—business model in the world: that of data mining and hyper-targeted advertising.[174]

Ben Franklin would be amazed at what his advertising innovation has wrought.

The Digital Advertising Ecospace

Given the global reach and intimidating bottom lines of the companies we're talking about in this book, their platforms have made advertising,

---

[172] Jeff Desjardins, "How the Tech Giants Make Their Billions," *Visual Capitalist*, March 29, 2019, https://www.nytimes.com/2019/01/30/technology/facebook-earnings-revenue-profit.html.

[173] Chaim Gartenberg, "What is Facebook? Just ask Mark Zuckerberg," *The Verge*, March 8, 2019, https://www.theverge.com/2019/3/8/18255269/facebook-mark-zuckerberg-definition-social-media-network-sharing-privacy.

[174] Rana Foroohar, *Don't Be Evil*. (New York, NY: Random House, 2019), p. 56.

perhaps counterintuitively, much more democratic. While in the past a small or even medium-sized business would have found it too much of a stretch to pay for a print ad in a major newspaper, taking out an ad on Google or Facebook, the platforms that carry 60 percent of U.S. digital advertising,[175] is really quite affordable.

For one thing, it's no longer necessary to foot the bill to travel to New York to take a meeting in a slick Madison Avenue office with people who look like Jon Hamm, Christina Hendricks, or Elisabeth Moss. Digital advertising is a much-less-suave affair than the days dramatized in *Mad Men*. That's because schmoozing for deals has morphed into an electronic bidding process handled by computers, and software programs automatically choose the winning bid for an ad space and factor in who should see the ad and where it will be placed in that person's view, in real time.

Let me walk you through the process.

First, of course, you have a particular product or service to sell. Let's pick two simple examples: maybe you make beautiful, hand-thrown pottery, or maybe you own a local lawn care service—and you decide you want to increase your customer base. In both of our examples, you have a brick-and-mortar base: your pottery studio serves as your work space, with a small area you've carved out for retail display; or you run your lawn care business out of a warehouse space in an industrial part of your town. Of course, you have a website, like everyone else these days: potential customers *can* find you. But how will they do that if they don't know where to look? You need to tell them.

So, you research your advertising options and decide, like so many others have before you, that placing ads on Facebook will be a good bet. You log onto Facebook, click on the Ads Manager function on your business's page, and start to create your first ad. The first thing Facebook is going to prompt you to do is set an ad budget. Digital advertising is awash in technical jargon that relates to each nanosecond an ad will appear

---

[175] Suzanne Vranica, "Google, Facebook and Amazon Gain as Coronavirus Reshapes Ad Spending," *Wall Street Journal*, December 1, 2020, https://www.wsj.com/articles/google-facebook-and-amazon-gain-as-coronavirus-reshapes-ad-spending-11606831201.

before a viewer's eyes. There's the *ad impression*, the act of downloading an ad to the target webpage, and *viewability*, which measures how many times the ad was actually seen on the site. An ad is considered *viewable* if 50 percent of its image appeared in the user's view for at least one second, or two seconds if it's a video ad.

There is also an ocean full of terms and acronyms to help you decide if all this viewing is paying off—and to decide how you are going to pay for, or bid, on your ad. You can analyze your ad's effectiveness by its cost-per-thousand-impressions (CPM), which measures how many pairs of eyes have been set on your ad for at least one second; you can analyze it by way of cost-per-like (CPL), which is, as you've no doubt guessed, the number of "likes" your post receives as it makes its way around the web; or you can analyze it on the basis of cost-per-click (CPC), which measures how much it costs you each time a potential customer clicks on the ad and makes his or her way to your landing page or your web page, or wherever it is you are driving your traffic. Facebook conveniently breaks down the CPC for you on your ads manager dashboard, and, depending on the expert you're listening to at any given moment, the average CPC will cost you somewhere between fifty cents and a dollar.

Ultimately, you're going to figure out what the cost actually boils down to, based on what it costs you to convert a click to a new customer. So you do a little more research and calculation. If you figure the cost to get twenty clicks is, at the high end, twenty dollars, and the average conversion rate for a Facebook ad click truly is 9 to 10 percent (as any number of online advertising "experts" will have you believe), then you can expect to gain two new customers for a twenty-dollar investment. Two customers who might sign a contract to have their lawn mown once a week all summer long, or who might each buy a nice pottery serving bowl—totally worth the twenty bucks. So you sign up for this minimal weekly budget for your first week of advertising.

The next thing Facebook will prompt you to do is to create your audience. Now is your opportunity to tap into all that personal information Facebook has stashed in its databanks and laser-target your customers. You think: who needs to have their lawns mown? Homeowners, of course.

So your first criteria is for Facebook to show your ad only to users who own their own homes. You also have the option of choosing to show your ad only to men, or only to women, or, of course, both. The man of the house mowing his suburban lawn on a Saturday morning might have been an American cliché at one point. But you're well aware that today lawn mowing is a genderless chore; so you click "both" and move on. Finally, because your business is necessarily a local one, you ask Facebook to show your ad only to those potential customers who own homes located in the same town as your business, or within a twenty-five-mile radius of your business's base.

A potter's options for creating an audience for an ad are both broader and more subjective. Maybe you've observed that most of your in-person buyers so far have been women, so your first criteria for Facebook is to show your ad to women only. Then you think a little deeper about who might buy artisan bowls of the sort you throw. Well, your pots are, at the end of the day, kitchen equipment, so you decide to target women who cook and bake. Your pots are also expensive, so you add a minimum income requirement for those to whom Facebook will serve your ad. When it comes to the geographic location you want to target, however, you've got the whole world at your fingertips. Really, you do. Unlike the owner of the lawn care service, whose customer base is limited by the practical consideration of how cost-effective it is to haul his lawn equipment how far, you can ship your wares anywhere on the globe. You wonder, though, how you're going to be able to handle fulfillment of the orders if your ad is wildly successful the world over. So you decide to test the waters by limiting the ad's service to the United States.

Finally, you're going to be directed to either create your ad, or to upload one you've had created. You can opt to have your marketing department or an outside marketing company design your ad, though it's also possible to create an original ad yourself right within Facebook's ad manager. If you decide to create your own ad, Facebook walks you through the process at this point with a series of prompts. Facebook's manager will prompt you to upload an image, multiple images, or a video—the last which you might consider making, as Facebook will tell you that, while images attract the attention of users as they're scrolling through their feeds, videos are even more of a viewer magnet. It will instruct you to input a headline, headline copy, and text in the appropriate places in their template. The quality of an ad that results from using the template can vary, depending on things like the resolution of the image you choose and the text you choose to pair it with. But the layout is going to be professional-looking and clean. Remember, the publisher of the web page, in this case Facebook, has a stake in the appeal of your ad: they want users to click on it, because they don't get paid unless that happens.

The real action starts when your ad is posted to the Facebook newsfeed and a user interacts with it. The phrase *programmatic advertising* refers to the process of the sale and placement of your ad into newsfeeds by algorithms. The algorithms sort through the characteristics you've indicated that you wanted of the people who are exposed to your ad, and they check on the going rate and *bid* on your ad's placement. By *bid*, I'm talking about the manner in which advertising inventory is bought and sold. The process is called *real-time bidding* (RTB), and it is done on a per-impression basis. In the social media world, "impression" is a metric used to determine how many times your ad has loaded from the server to a person's browser. The algorithms do all this, matching your ad at an appropriate rate with the people you want to see it, and they do it in a fraction of a second. Algorithms are *fast*.

Moving to Google: in the pay-per-click digital advertising world, the average cost-per-click for an ad that shows up on a Google search page is one to two dollars. For an ad that appears on websites connected to Google's ad program—which is virtually millions of sites, including YouTube—the cost is one dollar or less.[176] How do those one- or two-dollar clicks add up? Well, Google controls about 90 percent of the world's search engine market. More than 63,000 Google searches occur every second—that's about 5.4 billion per day. And Google-owned YouTube, poised to be a "media juggernaut" that could soon equal Netflix in rev-

---

[176] Ryan Maake, "How Much Does Google Ads Cost?", WebFX, (December 21, 2020), https://www.webfx.com/blog/marketing/much-cost-advertise-google-adwords/.

enue,[177] has over 1.9 billion users each month. Google's Gmail has more than 1.5 billion active users. More than 2.9 million companies use at least one of Google's digital marketing services.[178] The math is staggering.

Now, taking out an ad on Facebook, Google, or any Big Tech site that accepts advertisements—and, especially, bidding on the ad—is more complicated than I've indicated above. Whole books have been written about it, and this is not one of them. What I want you to take away from our discussion here is that what the algorithm is bidding on, on behalf of the advertiser, is *you*. All of this high-tech, high-speed, algorithmic advertising requires your personal data to create the targeted ads that millions of businesses rely upon.

But being able to target you with ads isn't the only value of your personal information to the businesses and organizations who've culled it from your online interactions. They also sell or share your data with their "commercial partners" and other companies.

For example, PayPal shares customer data, such as name, address, phone number, date of birth, and bank account information with its "partners," like Facebook, marketers and publicists including Google and LinkedIn, credit agencies like Experian and Equifax, and various banks.[179] It also shares customers' IP addresses—short for Internet Protocol address, the unique identifying number connected to each piece of hardware, such as your computer, that allows your device to communicate with other devices over an Internet Protocol-based network, like the internet.

The smart objects in our homes that interact with us by voice commands or an app also collect our personal data to keep improving their

---

[177] Jennifer Elias, "YouTube Is a Media Juggernaut that Could Soon Equal Netflix in Revenue," *CNBC* (April 27, 2021, https://www.cnbc.com/2021/04/27/youtube-could-soon-equal-netflix-in-revenue.html.

[178] Deyan Georgiev, "111+ Google Statistics and Facts that Reveal Everything About the Tech Giant," *Review42*, (November 21, 2020), https://review42.com/google-statistics-and-facts/.

[179] Wharton, "Your Data Is Shared and Sold…What's Being Done About It?", Knowledge @ Wharton, (October 28, 2019), https://knowledge.wharton.upenn.edu/article/data-shared-sold-whats-done/.

products, and they sell it to other companies as well. For example, a "seemingly innocuous" app, MyQ from Chamberlain, works with a forty-dollar hub that connects with a Wi-Fi router, so you can open and close your garage door remotely. The privacy label reveals: "The following data may be used to track you across apps and websites owned by other companies...." The list of information that *a garage door opener* collects from you includes your name, email address, "product interaction" information, and more, to target its customers with ads across the internet. And it also shares this data with Amazon.[180]

The Internet of Things (IoT) refers to the network of physical objects that contain or are embedded with software, sensors, and/or other technologies so that they can be connected to and exchange information with other devices and the internet system itself. The garage door device we just spoke about is part of the IoT, as is your smart TV. Smart TVs are an amazingly affordable technology, and their price point is low because the manufacturers are willing to make only a slim profit on them. The real money is in selling the data they mine while you're watching your streaming networks. Because high-tech hardware is expensive to make, these manufacturers are transforming their business models to trade in data.

"Popular brands of TV sets keep track of what we're watching and report it to companies that want to sell us new cars or credit cards," writes *New York Times* tech columnist Shira Ovide. "One reason they do it is that selling personal information is pure profit, whereas selling you a TV set is definitely not."[181] This is the same business model that computer printer manufacturers work with, selling you the actual printer for a low cost, but keeping the price of the ink you need for the printer to function high. Only selling your private information is so much more lucrative than ink.

---

[180] Brian X. Chen, "What We Learned from Apple's New Privacy Labels," *New York Times*, (January 27, 2021), https://www.nytimes.com/2021/01/27/technology/personaltech/apple-privacy-labels.html.

[181] Shira Ovide, "Why your TV spies on you," New York Times, (January 25, 2021), https://www.nytimes.com/2021/01/25/technology/why-your-tv-spies-on-you.html.

In 2019, a first-of-its-kind study exposed the flood of data that is collected from many of the popular smart devices we use in our homes, including smart TVs. Researchers from Northeastern University in Boston and Imperial College London ran a study about smart device privacy issues that involved 34,586 experiments to see what kind of data smart devices were collecting and sharing. The researchers watched data flow from home security cameras and doorbells, thermostats, light bulbs, voice-command audio assistants, refrigerators, washers and dryers, microwaves, and hubs that controlled multiple devices.

The study's findings confirmed that the majority of the online devices in our homes share our data with data-hungry businesses. An astounding seventy-two of the eighty-one Internet of Things devices being tested sent data to third-party companies not related to the manufacturer. Thirty of the eighty-one devices shared data without safeguarding it with any encryption, making it easy for eavesdroppers to intercept and uncover the user's identity, location, and behavior with the device. Many devices shared data with Amazon, Google, and Microsoft, which did not surprise the researchers because these companies provided the cloud services that connected the devices. They were vexed, however, to find that nearly all the TVs in their tests sent data to Netflix, even though they never configured any TV with a Netflix account. They watched data stream to Facebook and data-tracking companies. They tracked video doorbells sending videos to their service providers "without any notification or consent from recorded parties." They couldn't explain why video cameras began recording without any movement to trigger them, "which is both unexpected and potentially suspicious. We do not have a root cause for this behavior." The study concluded that most devices used encryption that protected users' personally identifiable information, but at the same time, an astounding 57.45 percent of the data collected went to third-party companies.[182]

---

[182] Jingjing Ren, Daniel J. Dubois, David Choffnes, Anna Maria Mandalari, Roman Kolcun, & Hamed Haddadi, "Information Exposure From Consumer IoT Devices: A Multidimensional, Network-Informed Measurement Approach," (October 21-23, 2019), Proceedings of the internet Measurement Conference, p. 267-279, https://moniotrlab.ccis.neu.edu/wp-content/uploads/2019/09/ren-imc19.pdf.

Another study from 2019 focused on two popular TV streaming devices, Roku and Amazon Fire TV, to uncover how they handle user data. In this first large-scale study of privacy issues with streaming channels, the researchers from Princeton University and the University of Chicago found that 69 percent of Roku channels and 89 percent of Amazon Fire TV channels sent data to services that collect data for digital advertising, such as SpotX, Google's DoubleClick, Nielsen Online, and Tremor Hub. As in the smart device study, these researchers concluded that data-collecting and sharing is widespread with these devices, and unique identifiers, such as device IDs and serial numbers, are sent over unsafe, unencrypted connections. In addition, the study revealed that it was nearly impossible for users to switch off the data-sharing. Why the lack of data privacy with these popular streaming products? The researchers minced no words: "The tendency of companies to monetize users' data makes it difficult to incentivize privacy friendly practices."[183]

Google and Facebook's advertising business model—built on monetizing the data that comes from our devices—is a boon to businesses that can use it to spend their advertising dollars efficiently. According to Google, the average business earns eight dollars in revenue for every one dollar spent in ads that appear on Google search pages. They also earn two dollars for every one dollar they spend on ads that go throughout other Google-owned websites, such as YouTube.[184] This digital ad ecosystem opens new doors for businesses, but let's be clear: it isn't only tech companies that collect and share your data. Even the American Civil Liberties Union (ACLU) also shares the data it culls—and sometimes

---

[183] Acar Mohajeri Moghaddam, "Watching You Watch: The Tracking Ecosystem of Over-the-Top TV Streaming Devices," Proceedings of the 2019 ACM SIGSAC Conference on Computer and Communications Security, pp. 131-147, https://www.princeton.edu/~pmittal/publications/tv-tracking-ccs19.pdf.

[184] Google, "Methodology," 2021, Google Economic Impact, https://economicimpact.google.com/methodology/.

they share it with Facebook, "one of the companies whose policies it often criticizes."[185]

Doing Biz as Apple, Amazon, and Microsoft

While selling advertising is the ground game for Google and Facebook, the other three giants of Big Tech—Amazon, Apple, and Microsoft—earn most of their billions selling products. Apple, the most valuable company in the world, with a market worth of two trillion dollars,[186] credits most of its success to sales of the iPhone, which accounts for about 55 percent of its sales. The iPad, MacBook, and iMac make up about another 20 percent of Apple revenue, followed by its iCloud and other services that—along with everything else in tech—have been seeing more growth in the age of COVID-19.[187] But that doesn't mean that users' personal data isn't also factored into Apple's business model. Apple collects your information from your devices and lets advertisers target you when you're using the App Store and News app. It may sell ad space on a minute scale compared to Google and Facebook, but the fact is Apple knows a lot about you and is storing and monetizing your personal data.[188]

Amazon, too, gets most of its revenue from its flagship online retail store that sells 12 million products, not including books or media,

---

185 Ina Fried, "ACLU Says It Shares User Data with Facebook, a Frequent Target of Criticism," AXIOS https://www.axios.com/aclu-data-shares-facebook-4f1d21f4-d432-4998-8c67-364500015c28.html.

186 Ben Popken, "Apple Is Now Worth $2 Trillion, Making it the Most Valuable Company in the World," *NBC*, (August 19, 2020), https://www.nbcnews.com/business/business-news/apple-now-worth-2-trillion-making-it-most-valuable-company-n1237287.

187 Shira Ovide, "How Big Tech Won the Pandemic," *New York Times* (April 30, 2021), https://www.nytimes.com/2021/04/30/technology/big-tech-pandemic.html.

188 Jefferson Graham, "Is Apple really better about privacy? Here's what we found out," *USA Today*, (April 17, 2018), https://www.usatoday.com/story/tech/talkingtech/2018/04/17/apple-make-simpler-download-your-privacy-data-year/521786002/.

directly, and acts as a marketplace for another roughly 340 million products.[189] Think of it this way: Amazon sells, on average, $4,722 in merchandise *per second*. That comes to around $238,000 in sales per minute, or $17 million in sales every single hour of every single day.[190]

In addition to being the world's second largest retailer[191]—WalMart is number one—Amazon is also the leader in cloud, or online, computing sales. We've touched on what a cloud service is in an earlier section, but I want to expand on that here because, today, businesses are commonly run online instead of through software that comes in a box, or through in-house computing systems. Cloud services do everything that in-the-building infrastructures used to do, such as create and store files, run applications, manage inventory, process email and video communications, and, of course, store data from customers. Employees can access all of these services from anywhere with any kind of device. Amazon Web Services (AWS) is the most popular cloud service by far, generating $13.5 billion in revenue in 2020 and making up "more than 63% of [Amazon's] operating profits for the year, on annual AWS revenue of $45.3 billion...."[192] In the first quarter of 2021, Amazon reported that AWS revenue grew by 32 percent.[193]

These are not, however, the only reasons for Amazon's clout. Amazon's incredible size has given it the power to discount its products and overtake whole sectors, such as book and music stores. I saw this impact first-hand

[189] Retail TouchPoints, "How Many Products Does Amazon Carry?", n.d., https://retailtouchpoints.com/resources/how-many-products-does-amazon-carry.

[190] "15 Amazon Statistics You Need to Know in 2021," *RepricerExpress*, https://www.repricerexpress.com/amazon-statistics/.

[191] Ty Haqqi, "5 Biggest Retailers in the World Heading Into 2021," *Insider Monkey* (December 24, 2020), https://www.insidermonkey.com/blog/5-biggest-retailers-in-the-world-heading-into-2021-911272/?singlepage=1.

[192] Todd Bishop, "Amazon Web Services Posts Record $13.5B in *Profits* for 202 in Andy Jassy's AWS Swan Song," *GeekWire* (February 2, 2021), https://www.geekwire.com/2021/amazon-web-services-posts-record-13-5b-profits-2020-andy-jassys-aws-swan-song/.

[193] Jordan Novet, "Amazon's Cloud Division Reports 32% Revenue Growth," *CNBC* (April 29, 2021), https://www.cnbc.com/2021/04/29/aws-earnings-q1-2021.html.

in just one neighborhood of New York City, when the Tower Records and Barnes & Noble stores near Lincoln Center closed in 2006 and 2011, respectively, and the Borders bookstore three blocks away at Columbus Circle also shut down in 2011. The fate of these stores followed nationally: every Tower store in the United States closed in 2006, and the entire Borders chain died in 2011.

Since Amazon opened its virtual doors as an online bookseller in 1994, the retail world has progressively shifted online and transformed the economic and physical landscape. Ecommerce brings convenience, and Amazon—as the gatekeeper of ecommerce, with roughly fifty percent[194] of all ecommerce sales—brings cheaper prices, all of which is great for the consumer. But the three stores in a four-block area of New York I mentioned above held more value than their earnings alone. They were destinations for tourists and other visitors, and part of the cultural fabric of the neighborhood. Ironically, Amazon recently entered the brick-and-mortar bookstore business it upended, opening its first store in Seattle in 2015. The first New York City store opened in 2017—in the same building that once housed the Borders store.

Amazon's impact on the retail industry is known as the "Amazon effect," describing shoppers' preference for the ease and convenience of online shopping and the resulting downturn or closure of physical stores. Some of the practices that account for the company's incredible growth have made it the target of lawmakers who say that Amazon holds "monopoly power," with the ability to derail competition by highlighting its own brands and undercutting competitors' prices as the in-house brands get a foothold in the marketplace. The clincher is, of course, that Amazon owns the playing field where the bulk of online buying and selling is done. If you have a product you want to sell, there is no bigger showroom than Amazon—a showroom wherein Amazon itself can push your products to the farthest, darkest corner in order to allow their own products to shine.

---

[194] "15 Amazon Statistics You Need to Know in 2020," ibid.

The same too-big-to-be-fair argument is also made against other Big Tech companies, including Google, which dominates internet searching and its advertising revenues with both the most popular browser, Chrome, and the most-used search engine, Google. One way the company achieved this was by making deals with other companies to ensure that Google would be the default search engine of choice.

Microsoft's largest earnings come from its Office products—Word, Excel, PowerPoint, Outlook, Access, Publisher, and OneNote—which are sold both as traditional software programs and cloud-based services. These products make up about one-quarter of the company's earnings, with another quarter coming from its cloud service, Azure, Microsoft's rival to Amazon's AWS. The next-largest chunk of the company's revenue comes from its Windows operating system, and 6.1 percent of Microsoft's earnings are from the ad space it sells on its Bing search engine.[195] Microsoft is a trillion-dollar company—one of the very few of those in the world; indeed, it is ranked as the fourth-biggest company in the world in terms of market value, and its co-founder Bill Gates is a household name, like the rest of Big Tech's marquee personalities.

Even though Microsoft, like Amazon and Apple, has a different business model than Google and Facebook, all five companies have faced the same scrutiny in Washington, D.C., over the power they exert in the tech economy—and this scrutiny isn't something new.

Microsoft was embroiled in a high-profile lawsuit brought by the Department of Justice back in the 1990s. This suit was an introduction for many of us to Big Tech's ability to stomp out competition. The Justice Department argued that Microsoft closed out opportunities for web browsers like then-existing services such as Netscape Navigator by illegally bundling its own web server, Internet Explorer, with its Windows operating system, which dominated the market. Microsoft was nearly broken up—indeed, for a time, the company was under a breakup order—but that order was later overturned on appeal and the litigants

---

[195] Jeff Desjardins, "How the tech Giants Make Their Billions," *Visual Capitalist*, (March 29, 2019), https://www.nytimes.com/2019/01/30/technology/facebook-earnings-revenue-profit.html.

reached a narrow settlement. While this settlement didn't determine if the bundling was illegal or not, nor did it require Microsoft to unbundle its browser, it did restrict the terms that Microsoft could impose on PC makers who distributed Windows.[196]

This more-than-twenty-year-old suit is quite similar to one brought in 2020 by the Department of Justice against Google. The more recent suit "focuses on how Google pays distributors, including mobile device makers and mobile carriers, to ensure they make its search engine the default."[197] There has been, as of this writing, no resolution to this suit, which seeks to "restrain Google LLC (Google) from unlawfully maintaining monopolies in the markets for general search services, search advertising, and general search text advertising in the United States through anticompetitive and exclusionary practices, and to remedy the effects of this conduct,"[198] though antitrust fervor is increasing in Washington, D.C., and on both sides of the ideological aisle. For example, in May, 2021, Senator Mike Lee (Utah) and Representative Ken Buck (Colorado), both Republicans, pressed Attorney General Merrick Garland to investigate potential anticompetitive behavior by Amazon in its pursuit of a ten-year, ten-billion-dollar cloud-computing contract with the Department of Defense (DoD) by "improperly influencing the Joint Enterprise Defense Infrastructure [JEDI] procurement procedure."[199]

Concern about the monopolistic behavior of the giants of Big Tech is nothing new. In 2019, the U.S. House of Representatives approached the issue in a more holistic manner and launched an investigation. Over the next sixteen months, the House Judiciary Committee's Subcommittee

---

[196] Matt Rosoff, " DOJ Case Against Google Has Strong Echoes of Microsoft Antitrust Case," *CNBC* (October 20,2020), https://www.cnbc.com/2020/10/20/doj-case-against-google-has-strong-echoes-of-microsoft-antitrust-case.html.

[197] Ibid.

[198] Case Document, https://www.justice.gov/opa/press-release/file/1328941/download.

[199] Dana Mattioli, "GOP Lawmakers Urge Probe of Amazon's Pursuit of Pentagon Contract, *Wall Street Journal* (May 3, 2021), https://www.wsj.com/articles/gop-lawmakers-urge-probe-of-amazons-pursuit-of-pentagon-contract-116201 00134?mod=djem10point.

on Antitrust, Commercial and Administrative Law held hearings that included a July 29, 2020, facedown with the leaders of four of the companies we're spotlighting in this book. For six hours, Amazon CEO Jeff Bezos, Google CEO Sundar Pichai, Apple CEO Tim Cook, and Facebook CEO Mark Zuckerberg fielded questions from committee members. According to the 490-page report the committee later released, "We pressed for answers about their business practices, including about evidence concerning the extent to which they have exploited, entrenched, and expanded their power over digital markets in anticompetitive and abusive ways." The CEOs' responses did not satisfy. "Their answers were often evasive and non-responsive, raising fresh questions about whether they believe they are beyond the reach of democratic oversight," the report stated.[200]

Nevertheless, the committee was able to draw profound and damning conclusions. "By controlling access to markets, these giants can pick winners and losers throughout our economy. They not only wield tremendous power, but they also abuse it by charging exorbitant fees, imposing oppressive contract terms, and extracting valuable data from the people and businesses that rely on them."[201] While acknowledging the benefits these once "scrappy, underdog startups"[202] have delivered to society, the Committee charged that "[t]hese firms typically run the marketplace while also competing in it—a position that enables them to write one set of rules for others, while they play by another...."[203] In short, the "hearings produced significant evidence that these firms wield their dominance in ways that erode entrepreneurship, degrade Americans' privacy online,

---

[200] U.S. House of Representatives Subcommittee on Antitrust, Commercial and Administrative Law of the Committee on the Judiciary, "Investigation of Competition in Digital Markets" (October 6, 2020), https://judiciary.house.gov/uploadedfiles/competition_in_digital_markets.pdf?utm_campaign=4493-519.

[201] Ibid.

[202] Ibid.

[203] Ibid.

and undermine the vibrancy of the free and diverse press. The result is less innovation, fewer choices for consumers, and a weakened democracy."[204]

The consequences of the behavior of Big Tech—their ongoing adherence to a flawed business model—are dire, and it bears repeating here that this business model is reliant on *you*. As soon as you create a log-on ID for a Google, Facebook, Amazon, Apple, or Microsoft account, these companies begin collecting information about how you engage with their platform. Their tracking then spreads to "off-platform" data—meaning data not collected directly through platforms owned by the company, but that the company observes thanks to its relationships with other businesses, from mobile apps to retail stores to data brokers, who package up data to sell to other companies.

As you continue to use these companies' products and online services, they gather more and more information about you, and this enhances their ability to improve their products and the amount of data they can turn into cash. Your benefit, however, remains fixed to one thing: your use of the platform or product. Dipayan Ghosh, the co-director of the Digital Platforms & Democracy Project at Harvard and former privacy advisor at Facebook, and Nick Couldry, a sociologist of media at the London School of Economics and Political Science, explain that this lopsided relationship is all wrong. "This is not how any economic relationship should be structured," they wrote in late 2020. "It should favor fair economic exchanges based on mutual understandings of the actual value at hand."[205]

Despite new efforts for data transparency from most tech companies, we're still far from fair economic exchanges with them. When Big Tech made data mining the new normal in doing business, it tore a gaping hole in this basic aspect of healthy economics. Business models that rely on improving products by mining personal information, and earning

---

[204] Ibid.
[205] Dipayan Ghosh and Nick Couldry, "Digital Realignment: Rebalancing Platform Economies from Corporation to Consumer," Harvard Kennedy School M-RCBG Association Working Paper Series No. 155, https://www.hks.harvard.edu/sites/default/files/centers/mrcbg/files/AWP_155_final2.pdf.

revenue from using that data in any way to make it valuable to someone else, is "rank exploitation" of our private information, according to Ghosh and Couldry. The model, however, rolls on and expands because the risks this data-mining poses, they write, are not usually foremost in our minds: "Users often fail to consider the full effect of data collection and monetization by firms—precisely because they fail to recognize the *future* value to them of keeping that information private (and the future harms—economic, social, psychological—that may result from failing to do so). When the typical customer signs up for a social media service, she does not usually consider that subscribing now might implicate the privacy of her data when exposed to a cybersecurity breach five years later—or perhaps even worse, when her vote is manipulated by Russian disinformation operators fifteen years later."[206]

This is the insidious new economic reality of our time—a hard, cold fact about surveillance capitalism and the billions that have been, and continue to be, reaped from our personal information.

Too Big to Tax

It's almost a cliché at this point to say that the debate about the tax rate corporations pay, and the amount of their dollars that actually make it to the U.S. Treasury, is a heated one. Focusing on Big Tech, it has been estimated that this industry alone has used legal loopholes in the tax code to avoid over $100 billion in taxes[207]—and/or inflated the amount they said they had paid in taxes by $100 billion over the period of a decade.[208] It's hard to know which statement is true, or if both are true, based on

---

[206] Ibid.

[207] Erik Sherman, "A New Report Claims Big Tech Companies Used Legal Loopholes to Avoid Over $100 Billion in Taxes. What Does that Mean for the Industry's Future?" *Fortune* (September 6, 2019), https://fortune.com/2019/12/06/big-tech-taxes-google-facebook-amazon-apple-netflix-microsoft/.

[208] Rupert Neate, "'Silicon Six' Tech Giants Accused of Inflating Tax Payments by Almost $100bn," Yahoo Finance (May 31, 2021), https://ca.finance.yahoo.com/news/silicon-six-tech-giants-accused-070116735.html.

what we know about the actual income of the companies in question. Let's break down those billions by first turning our attention to one example, featuring one company.

When Long Island City, New York, won the North American auction to be selected as Amazon's second headquarters in 2018, some New Yorkers were thrilled. For fourteen months, cities in the United States and Canada had wooed Amazon with incentives in hopes of landing the deal that Amazon pledged would bring 25,000 jobs and a building investment of $5 billion. Just over a year after bidding had begun, Amazon announced that the new headquarters would be split between New York and Arlington, Virginia, and New York Governor Andrew Cuomo and Mayor Bill de Blasio finalized the giveaways Amazon would receive: in exchange for building their headquarters in the city that runs along the East River in Queens, the governor and mayor promised $3 billion in economic incentives. Those deal sweeteners included tax breaks and $500 million to help build the headquarters.[209]

The New York City Council, which was not privy to the deal-making and learned only after the fact that the deal would bypass the city's land-use review, along with several citizen groups and public officials, pushed back on the incentives. In a statement, one city council member and state senator wrote, "Offering massive corporate welfare from scarce public resources to one of the wealthiest corporations in the world at a time of great need in our state is just wrong."[210]

Residents held protests supported by unions, elected officials, and advocacy groups; the city council ran heated public hearings; and opinions from both sides blazed through in the media. After three months

---

[209] Spencer Soper, Matt Day, and Henry Goldman, "Behind Amazon's HQ2 Fiasco: Jeff Bezos was Jealous of Elon Musk," *Bloomberg*, (February 3, 2020), https://finance.yahoo.com/news/behind-amazon-hq2-fiasco-jeff-100012951.html.

[210] Michael Gianaris, "Statement from Senator Michael Gianaris and Council Member Jimmy van Bramer Regarding Amazon LIC Deal," (November 11, 2018), https://www.nysenate.gov/newsroom/press-releases/michael-gianaris/statement-senator-michael-gianaris-and-council-member-0.

of this back-and-forth backlash, Amazon called it quits and withdrew its offer. "For Amazon, the commitment to build a new headquarters requires positive, collaborative relationships with state and local elected officials who will be supportive over the long-term," the company said on February 14, 2019.[211]

Mayor de Blasio criticized Amazon for not staying at the table and basically said good-riddance, writing, "You have to be tough to make it in New York City…. Instead of working with the community, Amazon threw away that opportunity." Governor Cuomo put the blame on the New York State Senate, saying that they had done "tremendous damage," prompting State Senator Michael Gianaris to respond that Amazon's behavior showed why "they would have been a bad partner for New York…. Rather than seriously engage with the community…Amazon continued its effort to shakedown governments to get its way."[212]

Now, I actually think that the creation of jobs is worth the giveaways in terms of tax breaks, particularly when you factor in the "multiplier effect." If a company brings 25,000 jobs to a city, that's 25,000 people who will be employed and who might otherwise not have found a job at all—and that's a valuable thing in and of itself.

But calculate the multiplier effect of those 25,000 new jobs. All of those newly employed people will need a way to get to work, and that means additional transit workers will have to be hired to drive and repair and clean the trains and buses that will get those new employees to and from their offices. All of those newly employed people will have to stock their pantries, and that means additional stock people and clerks will have to be hired at area grocers. Most of those people will have families, and that means to educate the workers' kids, local schools will have to hire more teachers and classroom aides and cafeteria workers and janitorial staff. Many of those people will want a cup of coffee to start the day, and that means area coffee shops and diners will have to hire

---

[211] Amy Plitt, "Amazon Cancels HQ2 Plans for New York City," *Curbed New York* (February 14, 2019), https://ny.curbed.com/2019/2/14/18224997/amazon-hq2-new-york-city-canceled.

[212] Ibid.

additional waiters and baristas. Each of the hypothetical 25,000 jobs a new company will bring to an area translates in this way to two, three, or even four additional jobs in order to provide the services those new employees will require.

The tension around this matter is, as you can see, nearly palpable. Where is that elusive line between attracting good jobs to an area, and the need to make corporations behave like good members of the community? The crux of the problem is neatly summed up by Axios's Ina Fried: "Attracting technology companies is a holy grail for economic development because they bring high-paying jobs and prestige to aspiring tech hubs. But that project is now colliding with some state leaders' efforts to rein in tech companies' growing power."[213]

A year later, however, Bloomberg News reported on the context of CEO Jeff Bezos's decision to create the large-scale bidding war and, when things got hot in New York, turn down the deal. Bezos had envied Tesla CEO Elon Musk for setting up a bidding war out West for a battery manufacturing plant that would create thousands of jobs. The five-state battle resulted in Nevada offering $1.3 billion in incentives for the winning bid. Bezos, who was receiving no government assistance from his company's home state of Washington, had also seen the state deliver an $8.7 billion handout to the Boeing company. To Bezos, the incentives Amazon was receiving for expansion projects around the country were trifling compared to what Musk and Boeing were getting, so he followed Musk's lead and launched the tax-incentive bidding war across North America for a second headquarters. Amazon insiders told the reporters that the new headquarters' negotiating team was convinced Amazon would be

---

[213] AXIOS Newsletter, https://web.archive.org/web/20210419135327/https://www.axios.com/newsletters/axios-login-36e066b9-4ef1-483f-b836-84a4e2a308a8.html.

welcome anywhere. The team's assumption had an aggressive edge to it, summarized behind closed doors as "F*** you. We're Amazon."[214]

That attitude sounds apropos in light of the massive power Amazon wields in the global economy. Big Tech companies have amassed so much power through their business models, acquisitions, expansions into multiple business types, and lack of regulation that they are acting as if they are governments themselves. Columbia University tech scholar and writer Alexis Wichowski coined the term "net state" to describe the new role Big Tech is carving out, venturing into activities that are normally the work of governments. Examples of these inroads include Tesla's energy venture in a handful of states,[215] and Microsoft's relationship with the United Nations.[216]

Microsoft's creation of a UN liaison office came three years after the world saw its first foreign ambassador to the tech industry. In 2017, Denmark created a new diplomatic post devoted to technology and named Casper Klynge ambassador. In Denmark's view, Big Tech companies had gained as much power as many governments, so it sent its new ambassador to Silicon Valley to represent its interests. Two years into his post, Klynge said, "Our values, our institutions, democracy, human rights, in my view, are being challenged right now because of the emergence of new technologies. These companies have moved from being companies with commercial interests to actually becoming de facto foreign policy actors." Denmark's foreign affairs minister added that "we've been too naïve for too long about the tech revolution" and needed

---

214 Ben Gilbert, "Amazon Had Cities Compete Over Its New Headquarters, and the Approach Was Reportedly Known Internally as, 'F--- you, We're Amazon.'" *Business Insider* (February 4, 2020), https://www.businessinsider.com/amazon-hq2-search-internal-motto-2020-2.

215 Gene Munster, "Tesla Is Becoming an Energy Company," *Loup Funds* (September 21, 2020), https://loupventures.com/tesla-is-becoming-an-energy-company/.

216 "Why Does Microsoft Have an Office at the UN? A Q&A with the Company's UN Lead," *Microsoft* (October 5, 2020), https://news.microsoft.com/on-the-issues/2020/10/05/un-affairs-lead-john-frank-unga/.

an ambassadorship to help "make sure that democratic governments set the boundaries for the tech industry and not the other way around."[217]

Wichowski interviewed Ambassador Klynge for her book, *The Information Trade: How Big Tech Conquers Countries, Challenges Our Rights, and Transforms Our World.* When asked for his view of technology, Klynge said, "The freight train is coming…. Technology will have a massive impact on international relations. It will have a massive impact on the convening power of the West. It will have a massive impact on the balance of power in the future."[218] Big Tech is equally aware of the fact. In 2020, Microsoft ramped up its association with global diplomacy by opening the aforementioned UN representation office in New York—and hiring Casper Klynge as its new vice president of European government affairs.[219] Klynge's switch to tech-*insider* "diplomat" is one more twist in the brave new world of the net state.

Elon Musk's Tesla entered the energy infrastructure business in 2017 in an alliance with Vermont's Green Mountain Power utility. The electric company offered customers the affordable sale or rental of Tesla "Powerwall" backup batteries as home generators attached to the electrical grid or their own solar array. Other states have seen the energy savings of Vermont's experiment, and Tesla's Powerwall is now integrated into some electrical grids in Connecticut, New York, Colorado, California, and Hawaii, as well as in the U.K. and Australia. "Tesla is now responsible for providing the infrastructure to light and heat the homes of over two million American citizens," Wichowski writes. "The question is whether those citizens are aware of this."[220] Residents may not give

---

217 Adam Satariano, "The World's First Ambassador to the Tech Industry," *New York Times* (September 3, 2019), https://news.microsoft.com/on-the-issues/2020/10/05/un-affairs-lead-john-frank-unga/.

218 Alexis Wichowski, *The Information Trade: How Big Tech Conquers Countries, Challenges Our Rights, and Transforms Our World,* HarperCollins (2020).

219 "Microsoft appoints senior government affairs leaders in Brussels and New York, establishes New York office to work with the United Nations," *Microsoft* (January 17, 2020), https://blogs.microsoft.com/eupolicy/2020/01/17/senior-gov-affairs-leaders-appointed-brussels-new-york/.

220 Wichowski, ibid.

much thought to who is bringing them energy, but Wichowski advises that they would be wise to be mindful that the service they think of as a public utility may actually be provided by a publicly owned corporation. Indeed, depending on the area where they live, it might *likely* be provided by a publicly owned corporation; there are any number of privately and publicly owned utility companies, and there are "public power utilities" as well. These last are community-owned, not-for-profit electric utilities that service some 49 million Americans in cities such as Seattle and Los Angeles, Nashville and Austin.[221]

This is not, in and of itself, necessarily a cause for concern, and can even come with upsides. Take the case of microgrids, which are local energy grids "with control capability, which means [they] can disconnect from the traditional grid and operate autonomously."[222] In cases where a storm causes an outage, or the traditional grid requires repair, a microgrid can take up the slack and continue to provide the electricity people need to heat or cool their homes, and to run their appliances—and it can often do so in a more locally controlled, cost-efficient, and environmentally friendly way.

So, then, why should we be vigilant about tech companies integrating with our public works at all? As Wichowski explains, Big Tech/net states do not carry the same obligations to the people as governments do. "Net states…differ from governments in two key ways. First, unlike governments, net states are under no legal obligation to maintain in perpetuity the infrastructure they implement…. Second, while net states do engage in contracts for services, they are not mandated by law to assist any other states or territories that should find their energy and telecom infrastructures wiped out by storms in the future…. The result of these two differences is that net state-sponsored aid may not last, or even be offered in the first place, regardless of the needs of the people.[223]

As Big Tech companies assume the power of governments, we can understand more clearly why Amazon took the stance it did toward cities

---

[221] Public Power website, https://www.publicpower.org/public-power.
[222] Energy.gov website, https://www.energy.gov/articles/how-microgrids-work.
[223] Ibid.

that dared question its demands. The world recently learned that these companies not only demand extraordinary incentives, but also minimize their tax burdens by the billions. In 2019, the British organization Fair Tax Mark researched the 2010 through 2019 U.S. tax filings of Facebook, Apple, Amazon, Netflix, Google, and Microsoft. They found a glaring discrepancy between the tax provisions companies set aside in their financial reports for paying taxes, and the amounts they actually paid, or "cash taxes." As I've already said, the combined gap of those unpaid taxes has been estimated to reach a shocking total of $100 billion.[224]

o o o

Perhaps the most egregious of all the illusions in Big Tech's bag of tricks are the illusions around the income they earn from selling our personal information, and the taxes they're paying on all the money they're earning on our backs. Indeed, reporting specific numbers concerning that income and those taxes is itself like trying to penetrate the tricks of the most sophisticated illusionist to find the answer to *how does he do that?*

Let's start with *exactly how much income does each of the Five derive from the advertisements they sell based on our data?* The overarching truth is that, while we can and do calculate informed estimates, there is no one, consistent way to get to a bottom-line number on this issue—or, really, any issue that involves the companies' finances. This is fundamentally because there is no consistent, across-the-board standard for reporting these numbers.

To start with, the companies honor different fiscal years. Facebook, Google, and Amazon adhere to a calendar year schedule, running January through December of any given year. Microsoft's fiscal year runs July to June, and Apple's is October to September. Now, it isn't unusual for companies to have different fiscal years. The problem is that these companies are intentionally opaque—they don't want us to know how much money

---

[224] Chloe Taylor, "Silicon Valley Giants Accused of Avoiding Over $100 Billion in Taxes Over the Last Decade," *CNBC* (December 2, 2019), https://www.cnbc.com/2019/12/02/silicon-valley-giants-accused-of-avoiding-100-billion-in-taxes.html.

they're actually making—and so they make the job of demystifying their bottom lines more difficult in a number of deliberate ways.

A major complicating factor is the categories into which income streams are aligned for any given company. For example, Google is owned by Alphabet, Inc., trades as both GOOG and GOOGL, and has about 160 subsidiary companies under its umbrella, including YouTube, FitBit, Waze, and Waymo, Google's entry in the self-driving car space. Microsoft reports "search advertising" revenue of $7.74 billion, but lists a separate $8 billion line item for LinkedIn revenue without mentioning advertising earnings[225] on that platform.[226]

Further, neither Amazon nor Apple provides stand-alone figures for ad revenue; instead they include them in larger, umbrella categories of "Other" and "Services", respectively. Amazon's "Other" category is "primarily" made up of "sales of advertising services, which are recognized as ads…delivered based on the number of clicks or impressions."[227]

Apple's category "Services" is even more confounding. "Services" net sales include: sales from the company's advertising, which includes various third-party licensing arrangements and the company's own advertising platforms; AppleCare, which is a portfolio of fee-based service and support products under the AppleCare® brand; digital content, which operates various platforms including the App Store® that allow customers to discover and download applications and digital content, such as: books, music, video, games and podcasts, as well as subscription-based services—including Apple ArcadeSM, a game subscription service; Apple Music®, which offers users a curated listening experience with on-demand radio stations; Apple News+SM, a subscription news and magazine service; Apple TV+SM, which offers exclusive original content; and Apple Fitness+SM, a personalized fitness service built for Apple Watch. Services

---

[225] This figure includes other income that Apple and Amazon include in "Other" and "Services".

[226] "Annual Report 2020," *Microsoft*, https://www.microsoft.com/investor/reports/ar20/index.html.

[227] "Amazon Annual Report," https://s2.q4cdn.com/299287126/files/doc_financials/2021/ar/Amazon-2020-Annual-Report.pdf.

net sales also include amortization of the deferred value of Maps, Siri, and free iCloud® storage and Apple TV+ services, which are bundled in the sales price of certain products—not to mention the payment services Apple Card™, a co-branded credit card, and Apple Pay®, a cashless payment service. Citing JP Morgan analyst Samik Chatterjee, a Reuters' article in late 2019 observed that "the company could leverage the millions of users who search its App Store and Safari browser daily to generate the stellar growth seen by Facebook and Google in recent years. He said the company had the potential to raise revenue by a third every year, from an estimated $2 billion currently to $11 billion in 2025." That's stunning growth by anyone's measure, though measurement is elusive: the article went on to point out that "Apple does not currently give detailed figures on advertising revenue."[228]

All that said, informed calculations based on the information they do disclose place total revenues of the Five, for their respective 2020 fiscal years, at around $304,054 billion. However, perhaps an even more accurate—and telling—way to look at the value of these companies is to refer to their *market caps*.

Market capitalization refers to how much a company is worth as determined by the stock market—the number of outstanding shares times the current market value of one share. A company's market cap will therefore change continually. I chose to use the date July 20, 2021 for my example; on that day the cap for all five companies was over $10.7 *trillion*.

A nation's gross domestic product (GDP) is one of the metrics used most frequently to measure a country's economic health. For the record, that combined total is much larger than, say, the annual GDP of the entire country of Japan.[229]

o o o

---

[228] "Apple Could Raise Annual Ad Income to $11 Billion by 2025: *JPMorgan*," Reuters (November 15, 2019), https://www.reuters.com/article/us-apple-advertising/apple-could-raise-annual-ad-income-to-11-billion-by-2025-jpmorgan-idUSKBN1XP17L.

[229] *The World Factbook*, CIA.gov website, https://www.cia.gov/the-world-factbook/countries/japan/#economy.

So, what do you think an annual tax bill ought to look like from companies that boast a collective value of $10.7 trillion? Would it help your calculations if I told you that Fair Tax Mark stated in a 2019 report that Amazon was expanding its global domination "on the back of revenues that are largely untaxed"?[230]

In 2020, the Five, collectively, set aside money earmarked for their tax provisions.[231] Tax provision is the amount a company sets aside in its financial reports to pay taxes, and cash taxes are the amounts that are actually handed over to the government.[232] In its financial reports, then, the Five set aside .03 percent of their combined market cap at year end for the purpose of paying taxes. Let's be clear: the Five set aside less than one half of 1 percent of their market caps for tax-paying purposes.

What's your tax rate? A little higher than one half of one percent? Indeed, the tax rate for an average American single filer earning less than $10,000 a year is 10 percent.[233] That rate rises to 12 percent for an individual earning up to $40,000; 22 percent for earnings up to $85,000; 24 percent for earnings up to $163,000; and 32 percent for earnings up to $207,000—all well over less than one half of one percent. But take a small step back with me: the federal poverty guideline for an individual is $12,880.[234] This means that, theoretically, an American citizen whose earnings fall below the poverty guideline is in a higher tax bracket than these multi-billion-dollar corporations. Let that sink in, and then take this into consideration too: the corporate tax rate in effect in 2020 was 21 percent.

But, let's be clear: taxes are paid on revenues, not total value. The next question, then, is, *what did the Five actually hand over to the government in*

---

[230] Fair Tax Mark, "The Silicon Six and their $100 billion global tax gap," December 2019, p. 6, https://fairtaxmark.net/wp-content/uploads/2019/12/Silicon-Six-Report-5-12-19.pdf.

[231] Approximately $33 billion.

[232] Taylor, ibid.

[233] Nerd Wallet website, https://www.nerdwallet.com/article/taxes/federal-income-tax-brackets.

[234] "2021 Poverty Guidelines," U.S. Department of Health and Human Services, https://aspe.hhs.gov/2021-poverty-guidelines.

hard cash? What were *their* tax rates? I predict you will not be surprised at all to hear that, once again, this is a hard figure to put one's finger on. This is primarily because while tax provisions represent all taxes to be paid—federal, state, local, and international—a "cash tax paid" notation in, say, an annual report often represents only federal taxes. Facebook and Apple reports were refreshingly forthcoming in this regard. Even if "cash tax paid" wasn't broken into categories, the bottom lines were straightforward: for 2020, Facebook listed $4.03 billion as its provision, and $4.23 billion as cash tax paid, with a tax rate of 12.2 percent;[235] and Apple listed $9.68 billion as its provision, and $9.501 billion as its cash tax paid, with a tax rate of 14.4 percent.[236]

Amazon, on the other hand, listed in its 2020 annual report $2.86 billion in tax provisions and $1.68 billion in cash taxes paid, net of refunds. According to the Institute on Taxation and Economic Policy, their "effective federal tax rate" was 9.4 percent in 2020—and 1.9 percent in 2019.[237] Microsoft, generally credited as taking the "least aggressive approach to tax avoidance,"[238] reported its tax provision for 2020 was $8.76 billion,[239] and a tax rate of 17 percent. Alphabet, the owner of Google, lists the taxes provisions for 2020 as $7.81 billion,[240] with an effective tax

[235] "Meta Platforms Income Taxes 2009–2021," *Macrotrends*, https://www.macro-trends.net/stocks/charts/FB/facebook/total-provision-income-taxes.

[236] "Apple Income Taxes 20062021," *Macrotrends*, https://www.macrotrends.net/stocks/charts/AAPL/apple/total-provision-income-taxes#:~:text=Apple%20in-come%20taxes%20for%20the%20twelve%20months%20ending,2019%20were%20%2410.481B%2C%20a%2021.62%25%20decline%20from%202018.

[237] "Amazon Has Record-Breaking Profits in 2020, Avoids $2.3 Billion in Federal Income Taxes, *Institute on Taxation and Economic Policy* (February 3, 2021), https://itep.org/amazon-has-record-breaking-profits-in-2020-avoids-2-3-billion-in-federal-income-taxes/.

[238] "The Silicon Six," Fair Tax Mark, https://fairtaxmark.net/wp-content/uploads/2019/12/Silicon-Six-Report-5-12-19.pdf.

[239] "Microsoft Income Taxes 2006–2021," *Macrotrends*, https://www.macrotrends.net/stocks/charts/MSFT/microsoft/total-provision-income-taxes.

[240] "Alphabet Income Taxes 2006–2021," *Macrotrends*, https://www.macrotrends.net/stocks/charts/GOOG/alphabet/total-provision-income-taxes.

rate of 16.2 percent, though how this figure breaks down over all 160 of Google's assets is less than transparent.

In short, not one of the companies came close to paying the full 21 percent corporate tax rate in effect that year.

A final word on taxes theoretically due versus taxes actually paid by the Five comes by way of the Organization of Economic Cooperation and Development (OECD). In 2019, it called for "a global minimum tax rate to curb large multinationals' tax avoidance" in response to the fact that "companies owning intangible assets often avoid paying taxes in countries where they make significant sales by declaring their profits in tax havens or registering their offices there."[241] At the G-7 in June of 2021, President Joe Biden supported the idea of a global tax rate.[242] This could have a significant impact on the way tech pays taxes and, of course, the amount they pay. The fear has long been that if the United States, where the Big Five tech companies are all based, raised taxes on tech to a more equitable level, then the rest of the world would make it easier for the companies to do business in their respective countries. A uniform tax would provide for a reasonable tax rate across borders, protecting the U.S. interests and ultimately, the interests of every other country in which tech does business.

And tech does business, literally, in every other country in the world.

Harkening back to our earlier example of companies such as Tesla entering the energy market, it seems wholly unwise to allow Big Tech to take unregulated control of sectors once the purview of governments— sectors such as energy and agriculture, which address our most basic needs. No government is perfect, but they are generally committed, by way of law, to guarantee continued maintenance of and access to such basic human services. Google, Facebook, Amazon, Apple, and

---

[241] "Tax Havens, A Paradise for US Tech Giants," *Berkeley Economic Review* (November 20, 2019), https://econreview.berkeley.edu/tax-havens-a-paradise-for-us-tech-giants/.

[242] David J. Lynch, "Biden set for G-7 boost in bid for all nations to impose minimum global corporate tax," *Washington Post* (June 1, 2021), https://www.washington-post.com/us-policy/2021/05/31/global-minimum-corporate-tax-biden-g7/.

Microsoft, on the other hand, are not. They are corporations, by definition committed to profit—entities that are neither patriotic to a country nor beholden to citizens who might depend on their services, should paying those taxes or providing those services interfere with their bottom lines. Tesla currently may be responsible for providing the infrastructure to light and heat the homes of over two million American citizens, but how durable would the commitment be to continue, should it become unprofitable to provide this service to those homes or any portion of those homes? How precarious would it be to rely on a corporation that is losing money for people who have become accustomed to enjoying heat and light in their homes?

○ ○ ○

When a magician's tricks are revealed to us, we are seldom disappointed. Instead, we're awed by the mechanics behind the illusion, the artistry of its execution, the flight of imagination that allowed the performer to dream up the stunt in the first place. If we think of the danger inherent in so many of these performances—everyone from Harry Houdini to Criss Angel has had their close calls—it's only because the risk increases the thrill. To dwell too intently on the death-defying nature of the tricks, however, would be a killjoy; we don't *want* to be disabused of our illusions.

We've lost a great deal in the past few years. The pandemic has shattered our sense of security and forced us to redefine what it feels like to live "normally." America's democratic crisis has shaken the faith of too many of our fellow citizens in the systems our forefathers and mothers created to sustain the nation. Even our technology, which proved during the course of 2020 to be more of a necessity than a convenience, has itself contributed to an overarching sense of anxiety—of having to face truths that were much more comfortable when the scales were still on our eyes. The machines that allowed us to do business, keep our children's education current, and stay in touch with one another in the loneliest days of lockdown were also, in large part, the source of false, misleading, and dangerous information—about COVID-19 vaccines, about the results of the 2020 presidential election. Now we've learned that it is our data—our most private information—upon which the massive fortunes of these tech companies have been built. And we've learned the extent to which tech companies manipulate the tax codes to avoid paying taxes on the enormous amount of money they make by sacrificing our privacy.

We have been disabused of so many of our illusions; systems seem to be failing all around us. In the United States, our trust in the tech sector this past year has dropped "precipitously, falling nine points, to an all-time low of 57."[243]

Well, hold on, because it gets worse. In the next chapter we're going to talk about the psychology of tech: the ways the Five adroitly manipulate us and impact our decision-making, and the scandals that expose the darkest side of data-mining. There are actions you can take to minimize encroachment on your own privacy, as well as ones we can take to have a greater impact on the global problem as a whole. But there's more disturbing information to process before we can get to the good news.

---

[243] Ina Fried, "1 big thing: Tech's public trust erodes," AXIOS (March 31, 2021), https://www.axios.com/newsletters/axios-login-faf55b82-d16f-4801-ae72-dc-3c307c34b1.html.

Chapter 3

Mind Games

James Williams is a former Google strategist who built the metrics system for the company's global search advertising business. These days he's a philosopher with a PhD from Oxford University. His expertise? Exploring the ethics of persuasive design. This is what he has to say about the industry in which he once toiled: it is the "largest, most standardised and most centralised form of attentional control in human history."[244]

The Big Tech business model we've been talking about has created what psychologist, economist, and Nobel Laureate Herbert A. Simon has dubbed the *attention economy*. Economics is a branch of learning concerned with the production, consumption, and transfer of resources; the *attention* economy refers to how the resource of human attention is allocated. In our internet era, when every bit of information collected over millennia by all of humankind is available with a few strokes of a finger on our smartphones, Simon suggests that "a wealth of information creates

---

[244] Paul Lewis, "'Our Minds Can Be Hijacked': the Tech Insiders Who Fear a Smartphone Dystopia," The Guardian (October 6, 2017), https://www.the guardian.com/technology/2017/oct/05/smartphone-addiction-silicon-valley-dystopia.

a poverty of attention."[245] If attention, then, is at a premium, and Big Tech's profits rely on our consumption of its contents—scrolling and liking and commenting and clicking links, and so forth—then Big Tech is going to respond to the attention deficit by doing everything in its power to keep our eyes on our screens as long as possible. Walk down almost any relatively populated street in almost any city or town in the world, and you will find pedestrians ignoring each other and often traffic, speaking into ear-mounted headphones or eyes glued to a mobile screen—all of which reveals how incredibly successful they have been at their game.

"I realized," Williams has said, "this environment of competition for people's attention was manageable in a pre-internet time, a time of information scarcity. But in a time of information overload, it's spilled into something qualitatively different. It's become an all-encompassing, persuasive environment."[246]

The science of grabbing our online attention and keeping it has its roots in a book written by B.J. Fogg, a Stanford psychologist, that launched a thousand tech engineers: *Persuasive Technology: Using Computers to Change What We Think and Do.* If that title doesn't make you wince, a few details of what Fogg called "captology"—the term of art for computers as persuasive technologies with the power to change people's minds, attitudes, and behaviors—just might.

In his research, writing, and university teaching, Fogg emphasized the positive applications of persuasive tech, such as websites that motivate people to quit smoking, manage their budgets, and reconnect with high-school friends. Because he considered such things as software bundles that control which search engine you use as clearly unethical coercion, and banner ads disguised as alarms about computer problems as pure

---

245 "Paying Attention: The Attention Economy," *Berkeley Economic Review* (March 31, 2020), https://econreview.berkeley.edu/paying-attention-the-attention-economy/.

246 Shona Ghosh, "A Former Google Strategist Says Tech is Warping Our Attention Spans––and It's Terrible for Humanity," *Business Insider* (May 31, 2018), https://www.businessinsider.in/a-former-google-strategist-says-tech-is-warping-our-attention-spans-and-its-terrible-for-humanity/articleshow/64405060.cms.

deception, Fogg probably did not envision that some of his students would apply captology to create apps, platforms, and devices that were addictive. Regardless of his intentions, Fogg's classes gave the future designers at Google, Facebook, Instagram, and elsewhere the foundation for building the attention economy in all its addiction-ridden glory. He claimed that designers and engineers who understand the ethical issues around tech persuasion "will be in a better position to create and sell ethical persuasive technology products,"[247] but some of his students must have been absent that day. In a report focused on persuasive communication, the National Academy of Sciences of the United States underscored the problem: "On the one hand, this form of psychological mass persuasion could be used to help people make better decisions and lead healthier and happier lives. On the other hand, it could be used to covertly exploit weaknesses in their character and persuade them to take action against their own best interest, highlighting the potential need for policy interventions."[248] Let's take a more detailed look at both hands.

In Fogg's model of persuasion, three things must occur at the same time to propel someone to do something: motivation, ability, and a prompt or trigger.[249] This is, actually, an easily understood equation. It's three in the afternoon, you're running to your fourth meeting of the day, you're tired, and you know you could use a pick-me-up. So when you spot a coffee kiosk in the lobby of the building where your meeting is being held, you fish your card out of your purse and buy yourself a quick cup. You want to be alert for your meeting, so you're motivated; you have a credit or a debit card on you to purchase some caffeine, so you have ability; you spot the kiosk and there's your trigger.

Fogg codified this simple equation, and Silicon Valley engineers ran with it to make their products addictive and, therefore, to increase users' highly profitable screen time. For example, social media such as Facebook, Twitter, Instagram, and Snapchat are designed to satisfy a

---

[247] B. J. Fogg, "Persuasive Technology: Using Computers to Change What We Think and Do," *Elsevier Science*, p. 235.

[248] Ghosh, ibid.

[249] Fogg Behavior Model website, https://behaviormodel.org/.

user's motivation in the form of their craving for social connection, or to allay their fear of social rejection. Theoretically, you can have thousands of "friends" on Facebook who will see the photo of you sitting on the beach and sipping a margarita on your vacation, and many of them will like the photo, or even comment on it—"Have another one for me!"—rewarding you with a nice little dopamine rush with each new bit of attention. Here's the hard question, though: how many of the people behind the likes and comments are your friends in real life or, as the kids say, IRL? How many of them do you meet for a drink after work? How many of them would you invite to your kid's wedding? How many of them would you recognize if you saw them on the street? ·

The *Social Brain Hypothesis* explains the correlation between the bonded relationships that characterize primate societies and the size of the primate neocortex. That is, in comparison to other animals, primates have large brains for their body size, and their societies are so much more complex than that of other species. It is their brain power that explains why they put so much time and effort into their interrelationships with other primates by, for example, grooming other members of their troop.[250]

Psychologist and anthropologist Robin I.M. Dunbar applied the hypothesis to humans. "We also had humans in our data set so it occurred to me to look to see what size group that relationship might predict for humans." What Dunbar found, according to Maria Konnikova, writing in the *New Yorker*, was, "judging from the size of an average human brain, the number of people the average person could have in her social group was a hundred and fifty. Anything beyond that would be too complicated to handle at optimal processing levels."[251] And Dunbar didn't stop there; he broke our human circles down into even more detailed layers. A person's

---

[250] Robin I.M. Dunbar, "The Social Brain Hypothesis and Human Evolution," *Oxford Research Encyclopedias* (March 3, 2016), https://oxfordre.com/psychology/view/10.1093/acrefore/9780190236557.001.0001/acrefore-9780190236557-e-44.

[251] Maria Konnikova, "The Limits of Friendship," *The New Yorker* (October 7, 2014), https://www.newyorker.com/science/maria-konnikova/social-media-affect-math-dunbar-number-friendships.

tightest circle includes just five "loved ones"; the next layer included fifteen "good friends"; then fifty "friends", 150 "meaningful relationships"; 500 "acquaintances", and 1500 "people you can recognize".[252] Indeed, Dunbar found that 1,500 was the very outside limit of how many people a person might be able to match a face with a name.[253] What this all boils down to for our purposes is that, while you might get a dopamine kick from having 3,000 "friends" on Facebook, you wouldn't be able to call half of them by their names if you ran into them in person.

Video games, on the other hand, break down Fogg's equation by tapping into the users' *motivation* to gain skills and feel a sense of accomplishment. The *ability* element means that there must be easy access to the games, in this case by making sure apps and platforms are simple to use. The third part of the change-your-behavior formula for persuasion—the prompt or trigger—comes in the form of videos that grab your attention, the rewards you receive for using an app or game, and your conscience, which compels you to immediately respond to notifications and messages so you won't be perceived as impolite or rejecting the person on the other end.

Sean Parker, the founding president of Facebook, has been up front about Facebook's exploitation of the attention economy, and our human psychological needs that foster it. "The thought process that went into building these applications, Facebook being the first of them…was all about: 'How do we consume as much of your time and conscious attention as possible?'…. And that means that we need to sort of give you a little dopamine hit every once in a while, because someone liked or commented on a photo or a post or whatever. And that's going to get you to contribute more content, and that's going to get you…more likes and comments…. It's a social-validation feedback loop…exactly the kind of thing that a hacker like myself would come up with, because you're exploiting a vulnerability in human psychology…. The inventors, creators—it's me, it's Mark [Zuckerberg], it's Kevin Systrom on

---

[252] "Dunbar's Number: Why We Can Only Maintain 150 Relationships," *BBC* (October 1, 2019), https://www.bbc.com/future/article/20191001-dunbars-number-why-we-can-only-maintain-150-relationships.

[253] Ibid.

Instagram, it's all of these people—understood this consciously. And we did it anyway."[254]

This addictive element of digital life is taking a toll on an entire generation, as uncovered by psychologist Jean M. Twenge of San Diego State University. "Teens who spend more time than average on screen activities are more likely to be unhappy, and those who spend more time than average on non-screen activities are more likely to be happy. There's not a single exception…. The more time teens spend looking at screens, the more likely they are to report symptoms of depression."[255]

Twenge calls the generation born between 1995 and 2012, who grew up with cellphones, "iGen", and notes that girls suffer more negative effects than boys because they use social media more often than do their male peers. Girls can, for example, feel more left out than boys because they more often see on Snapchat, Instagram, and Facebook posts showing their friends and classmates spending time together without them. Girls are also, unsurprisingly then, getting more depressed than boys: from 2012 to 2015, boys' symptoms of depression increased by 21 percent while girls' symptoms increased by 50 percent. Twenge writes, however, that suicide rates are increasing for both girls and boys, with part of the rise for girls due to their vulnerability to cyberbullying. "Social media give middle- and high-school girls a platform on which to carry out the style of aggression they favor, ostracizing and excluding other girls around the clock."[256] Indeed, according to the Centers for Disease Control, teen

[254] Mike Allen, "Sean Parker unloads on Facebook: 'God Only Knows What It's Doing to Our Children's brains," AXIOS (November 9, 2017), https://www.ax-ios.com/sean-parker-unloads-on-facebook-god-only-knows-what-its-doing-to-our-childrens-brains-1513306792-f855e7b4-4e99-4d60-8d51-2775559c2671.html.

[255] Jean M. Twenge, "Have Smartphones Destroyed a Generation?' *The Atlantic* (September 2017), https://www.theatlantic.com/magazine/archive/2017/09/has-the-smartphone-destroyed-a-generation/534198/.

[256] Ibid.

suicides increased by a shocking 57 percent in the decade between 2007 and 2017.[257]

While girls are often motivated to log on for social reasons, boys gravitate to online video games because they have "a developmental drive to gain abilities and accomplishments," according to psychologist Richard Freed. He explained that video games, designed with the persuasive formula that rewards players with cash boxes, coins, and rewards, make boys "feel like they are mastering something" and "creates bad [gaming] habits and statistically poor academic performance."[258] The overall impact of growing up with online access is seen in the trend of teenagers shifting away from a development norm: the desire for independence. Steady downturns since 2000 show that teens spend less time going out with friends, date less, have less sex, and are in no rush to learn how to drive.[259]

Lurking in the shadow of these realities about addictive tech is Facebook's admission that it knows how to identify teens' emotional states based on their behavior while on the site. The company profits from this information, telling advertisers that it can track teens feeling stressed, useless, worthless, and insecure, and thus pinpoint "moments when young people need a confidence boost." At the same time, however, Facebook denies that it provides tools "to target people based on their emotional state."[260] In other words, as an addictive platform that increases teen isolation and bullying, Facebook supplies the bullet, if not the gun.

Scientists, psychologists, and physicians have seen evidence of addiction in tech users of all ages, of course, because brains of all ages are all susceptible to the pleasure rushes that engineers design to draw and keep the attention of digital users. The so-called "pleasure center" of the

---

257 Sally C. Curtin, M.A., "State Suicide Rates Among Adolescents and young Adults Aged 10–24: United States, 2000–20018," *Center for Disease Control* (September 11, 2020), https://www.cdc.gov/nchs/data/nvsr/nvsr69/nvsr-69-11-508.pdf.

258 Chavie Lieber, "Tech companies use 'persuasive design' to get us hooked. Psychologists say it's unethical," *Vox*, August 8, 2018, https://www.vox.com/2018/8/8/17664580/persuasive-technology-psychology.

259 Twenge, ibid.

260 Ibid.

brain was discovered in 1954 by James Olds, an American psychologist, and Peter Milner, then a postdoctoral fellow at McGill University.[261] Dopamine, a neurotransmitter that stimulates contentment, happiness, and excitement, is released in this pleasure center (technically, the *nucleus accumbens*) as the result of some activity the human considers pleasurable: getting an "A" on a final exam; passing a driver's test; being the winning candidate in a job search; getting a raise or a bonus at work; finding a bargain while shopping; drinking alcohol; eating certain foods, and that includes sugar; having sex; gambling; using drugs, from cannabis to opioids. It doesn't matter if the activity is a positive one, such as closing on one's first house, or a negative one, like smoking crack; the brain reacts by rewarding the body with a shot of dopamine. And it is this shot of that neurotransmitter—and the accompanying feeling of contentment, happiness, or excitement—that plays a critical role in reinforcing, or rewarding, the activity. The continuing desire or need for more of this reward—the ongoing quest for that next shot of dopamine—is what reinforces addictive behavior. In addition to triggering engineered behavior, smartphones, for example, are addictive hardware devices in and of themselves. The distinct glow of the smartphone screen generates chemical reactions in the reward pathway of the brain, acting in the same way as does sugar or cocaine.[262]

Children's addictions to smartphones and other tech are especially concerning because their brains are still developing, and this development—including the shaping of their emotional needs and drives—can be severely impacted by addictive behavior. We often talk of our brains as our "gray matter," but the truth is that the brain's white matter, or *myelin*, is just as critical to efficient brain function. Myelin is a whitish substance made up of a mixture of proteins and phospholipids that forms

---

[261] Christopher Bergland, "The Neuroscience of Pleasure and Addiction," *Psychology Today* (May 31, 2014), https://www.psychologytoday.com/us/blog/the-athletes-way/201405/the-neuroscience-pleasure-and-addiction.

[262] Tanya Basu, "Just How Bad Is Kids' Smartphone Addiction?," *Daily Beast*, January 9, 2018, https://neuroscience.stanford.edu/news/just-how-bad-kids-smartphone-addiction.

an insulating, protective sheath around the brain's nerve endings. Among other factors, the myelin facilitates the speed with which brain functions are carried out. The development of the myelin sheaths, also known as *myelination*, happens over the course of the first nearly twenty-five years of a person's life.[263] The most pronounced myelination happens in the first two years after birth. It takes an additional ten or twelve years until most of the general construction of the brain is complete, with the "fine tuning" of the brain continuing until a person reaches nearly twenty-five years of age. The purposeful imposition of addictive behaviors upon not-yet-fully-developed brains can result in physical problems like poor sleep habits and poor stress regulation, psychological problems such as depression and ADHD-like symptoms, and it can be linked to decreased academic performance, poor social coping skills, and antisocial behavior.[264]

"The design of modern technologies is purposefully habit-forming and programmed with the sort of variable rewards that keep humans engaged," said Jenny Radesky, a pediatrician at the University of Michigan and expert on children's tech use. She avoids using the word "addiction," however, because that puts the blame on the tech user rather than "the digital environment that is shaping the individual's behavior, often through methods that are intentionally exploitative or subconscious."[265] This is a critical point, because it's easy to be annoyed or angry at the teen—or the husband or the wife—for being glued to their phone while sitting at the breakfast table, in a restaurant, or while visiting relatives, rather than interacting and being present in the moment. Instead, our judgment and frustration should be directed at the companies that deliberately create the addictive experience to make money.

---

263 "Myelination in Development," UCLA website, http://cogweb.ucla.edu/CogSci/Myelinate.html.

264 Gadi Lissak, "Adverse Physiological and Psychological Effects of Screen Time on Children and Adolescents: Literature Review and Case Study," *National Institute of Health: National Library of Medicine* (February 27, 2018), https://pubmed.ncbi.nlm.nih.gov/29499467/.

265 Perri Klass, "Is 'Digital Addiction' a Real Threat to Kids?," *New York Times,* (May 20, 2019), https://www.nytimes.com/2019/05/20/well/family/is-digital-addiction-a-real-threat-to-kids.html.

It helps to place into perspective the damage we may be inflicting unwittingly upon ourselves and our families when we take a bird's-eye view of how deeply technology is embedded in our lives:

- Typical smartphone users check their phones every twelve minutes from morning to bedtime.[266]
- Young people under age twenty-one check their phones every 8.6 minutes.[267]
- Children ages eight to eighteen spend an average of 44.5 hours per week in front of digital screens.[268]
- Nearly half of all two- to four-year-olds have played video games.[269]
- By age eight, 90 percent of children have used a computer, 81 percent have played video games, and 60 percent have used apps or played games on mobile tech such as a cell phone or tablet.[270]
- 45 percent of teens report that they use the internet "almost constantly."[271]
- 72 percent of teens and 48 percent of parents feel the need to immediately respond to digital notifications, including texts and Facebook's Messenger.[272]

---

[266] Kyle Pearce, "100 Mind Blowing Smartphone Addiction Statistics (2021) Update," *DIY Genius* (June 5, 2021), https://www.diygenius.com/smartphone-addiction-factsheet/.

[267] Ibid.

[268] "Internet Addiction Disorder––What Can Parents Do for Their Child?" *Webroot,* https://www.webroot.com/us/en/resources/tips-articles/internet-addiction-what-can-parents-do.

[269] Dr. Brent Conrad, "Media Statistics––Children's Use of TV, Internet, and Video Games," *Tech Addiction,* http://www.techaddiction.ca/media-statistics.html.

[270] Ibid.

[271] Katie Hurley, LCSW, "Addiction: Are You Worried About Your Child?" *PSYCOM,* https://www.psycom.net/cell-phone-internet-addiction.

[272] Ibid.

- 40 percent of adults, including 65 percent of those under age thirty-five, check their phones within five minutes of waking up.[273]
- The average smartphone user taps, swipes, or clicks on their smartphone 2,617 times a day.[274]
- 72 percent of regular smartphone users say they will feel anxious if they move five feet away from their phone.[275]

Indeed, we spend so much time online that we've created a new medical condition for ourselves: DES, or digital eye strain. "Nearly 60% of Americans experience some symptoms of DES."[276] Symptoms include dry eyes, a burning or stinging sensation of the eyes, or an inability to focus the eyes, and they can be the result of too much time in front of a screen. To avoid DES, the medical community suggests doing things like resting the eyes by looking away from the screen for at least twenty seconds every twenty minutes, avoiding overhead lighting to reduce glare on the screen, and using lubricant eyedrops to combat the symptom of dryness.

A whole mini-industry has arisen around the DES problem. The manufacturers of "blue light protection glasses" claim their eyeglasses can relieve DES, improve our sleep, and even prevent the development of macular degeneration which, they say, near-constant exposure to the blue light emitted from our devices can accelerate. These are big, bold claims, but unfortunately the science behind them isn't yet proven. According to a study published by the American Journal of Ophthalmology, "Blue-blocking lenses did not alter signs or symptoms of eye strain with computer use relative to standard clear lenses."[277] Until the science solidifies one way or the other around the effects of blue light on digital users,

[273] Pearce, ibid.
[274] Ibid.
[275] Ibid.
[276] "Eye Health in the Digital Age: Does Too Much Screen Time Hurt Your Vision?" *UAB Medicine News*, https://www.uabmedicine.org/-/eye-health-in-the-digital-age-does-too-much-screen-time-hurt-your-vision-.
[277] Sumeer Singh, Laura E. Downie, Andrew J. Anderson, "Do Blue-blocking Lenses Reduce Eye Strain From Extended Screen Time," (February 11, 2021), https://www.ajo.com/article/S0002-9394(21)00072-6/pdf.

should we consider the claims about blue light protection lenses as only another bit of disinformation? Personally, I don't think we should. I wear blue light lenses, and I find them helpful; but if you don't, perhaps you'd use a stronger word than "disinformation." After all, isn't disinformation often deployed as a method of persuasion? Let's state the sales pitch in another, more forthright manner: *You might go blind if you aren't persuaded to buy our blue lens glasses.*

We've talked about the persuasion, as well as the coercion, around the nuanced engineering that goes into creating addictive tech, and tech openly admits to doing both in their quest for our attention and their advertisers' dollars. We're in on the open secret. From our earliest days, we're conditioned to being presented with the latest colorful toy or newest sugary cereal along with the Saturday morning cartoons. If we use TV ratings as an indicator, we can say that a great many of us have watched *Mad Men*, which was, in one sense, an articulation of how our emotions and desires are vulnerabilities that make us easy marks for the invisible influence of advertisers.

Each person harbors specific hopes, fears, and anxieties, as well as physical and financial realities that can be tapped to influence our decision-making.[278] *Manipulation* is the process of secretly tempting, seducing, guilting, or otherwise playing upon those hopes, fears, and anxieties in a covert attempt to benefit the manipulators. Old-time advertisers told us that if women only smelled like Love's Fresh Lemons, we'd find everlasting love, and backed it up with the then-hit song by British singer-songwriter Donovan, "Wear Your Love Like Heaven."[279] Or if we all only voted for Lyndon Johnson, we wouldn't have cause to worry about a nuclear Armageddon.[280]

Online manipulation is much the same as the old-fashioned kind used to be, with digital tech now being used to "covertly influence another

---

[278] Ibid.

[279] "Love's Fresh Lemon TV Commercial—1970," https://www.youtube.com/watch?v=nYdHa5XWxVc.

[280] "'Daisy' Ad (1964): Preserved from 35mm in the Tony Schwartz Collection," https://www.youtube.com/watch?v=riDypP1KfOU.

person's decision-making by targeting and exploiting decision-making vulnerabilities."[281] The difference is that online manipulation can be accomplished much more precisely. Because advertisers now have so much personal information about us, they no longer have to create one ad, throw it out to the general public, and hope that it reaches some percentage of their target market. Now they can laser-target the lovelorn as well as the independent or one-issue voter, the man who needs a new sofa in his living room, and the woman who enjoys cashew chicken salad for lunch.

For example, placing a high-energy, girl-on-the-dance-floor beauty product ad in the Facebook newsfeeds of users who appear to be extroverts will elicit higher click-through rates (CTRs) and conversion rates (which are clicks that turn into purchases or downloads) than sending those extroverts a calm, subdued ad. A 2017 study ran this type of experiment on 3.5 million Facebook users. The researchers analyzed a database of the users' Facebook "likes" to identify traits that revealed them as introverts, extroverts, or people with low openness or high openness. High-openness people are curious, unconventional, individualistic, and imaginative. Low-openness people, on the other hand, prefer the familiar rather than the new or unusual and are more conservative and traditional. The researchers designed separate ads to appeal to each of the four traits and found a direct link between personality-matching and higher responses to an ad. Indeed, "matched" ads—that is, ads that are designed for people who exhibit certain traits and then targeted directly to them—resulted in up to 40 percent more clicks and up to 50 percent more purchases than mismatched or non-targeted ads. The study reported that, at the end of the day, "psychological targeting makes it possible to

---

[281] Daniel Susser, Beate Roessler, and Helen Nissenbaum, "Technology, autonomy, and manipulation," *Internet Policy Review*, vol. 8, issue 2 (2019), doi:10.14763/2019.2.1410. https://policyreview.info/articles/analysis/technology-autonomy-and-manipulation.

influence the behavior of large groups of people by tailoring persuasive appeals to the psychological needs of the target audiences."[282]

o o o

Before we can move on, however, we need to emphasize a few critical points.

First, psychological manipulation is as old as the hills; the precision of the targeting is what is new. Second, while the words misinformation and disinformation may seem to have taken on new and more sinister meanings in our contemporary culture, the concepts have a long history, not just in advertising but in politics. What might come first to mind when we think of historical instances of the peddling disinformation are the snake oil salesmen of the late 1800s, a "phrase [that] conjures up images of seedy profiteers trying to exploit an unsuspecting public by selling it fake cures."[283] We'd be remiss, however, to dismiss the term so lightly.

"Around 31 B.C., Octavian, a Roman military official, launched a smear campaign against his political enemy, Mark Antony. This effort used, as one writer put it, 'short, sharp slogans written on coins in the style of archaic Tweets.' His campaign was built around the point that Antony was a soldier gone awry: a philanderer, a womanizer and a drunk not fit to hold office. It worked. Octavian, not Antony, became the first Roman emperor, taking the name Augustus Caesar."[284] There really is nothing new under the sun if we can trace "fake news" as far back at 31 B.C.

---

[282] S. C. Matz, M. Kosinski, G. Nave, and D. J. Stillwell, "Psychological Targeting As an Effective Approach to Digital Mass Persuasion," Proceedings of the National Academy of Sciences of the United States of America, (November 28, 2017). DSM-5 info: https://www.ncbi.nlm.nih.gov/pmc/articles/PMC6876823/.

[283] Lakshmi Gandhi, "A History Of 'Snake Oil Salesmen,'" https://www.npr.org/sections/codeswitch/2013/08/26/215761377/a-history-of-snake-oil-salesmen.

[284] Michael J. O'Brien and Izzat Alsmadi, "Misinformation, Disinformation and Hoaxes," Salon (May 2, 2021), https://www.salon.com/2021/05/02/misinformation-disinformation-and-hoaxes-whats-the-difference_partner/.

We also need to take into account that gullibility can simply be part of human nature and plays its part when fake news goes awry—or goes viral. In two famous cases, events that were *supposed* to be hoaxes were taken as gospel by a credulous public.

The first is what became known as the Great Moon Hoax, when, in 1834, the English astronomer Sir John Herschel traveled to the Cape of Good Hope to catalog the stars in the Southern Hemisphere. The next year, 1835, the *New York Sun* published a six-part series of stories that purported to be about what Sir John discovered by looking through his telescope in that remote location. In the first installment, the author claimed that the astronomer had created a gargantuan telescope for the trip while still in England—a mechanical beast twenty-four feet in diameter and weighing in at seven tons—and transported it to South Africa where its colossal powers of magnification allowed Sir John to prove that there really was life on the moon. "First there were hints of vegetation, along with a beach of white sand and a chain of slender pyramids. Herds of brown quadrupeds, similar to bison, were found in the shade of some woods. And in a valley were single-horned goats the bluish color of lead…. But the real surprise came on day four: creatures that looked like humans were about four feet tall—and had wings and could fly. 'We scientifically denominated them as Vespertilio-homo, or man-bat; and they are doubtless innocent and happy creatures,'"[285] the author asserted—and readers believed. Richard Adams Locke, after confessing that he was the author of the series, "said that it was meant as a satire reflecting on the influence that religion had then on science. But readers lapped up the tale, which was soon reprinted in papers across Europe. An Italian publication even included beautiful lithographs detailing what Herschel had discovered."[286]

The second of these events took place on the evening of October 30, 1938, when a twenty-three-year-old Orson Welles and his *Mercury*

---

[285] Sarah Zielinski, "The Great Moon Hoax Was Simply a Sign of Its Time," *Smithsonian Magazine* (July 2, 2015), https://www.smithsonianmag.com/smithsonian-institution/great-moon-hoax-was-simply-sign-its-time-180955761/.
[286] Ibid.

Theatre on the Air performed a radio adaptation of H.G. Wells's forty-year-old novel *The War of the Worlds*, in which the narrative was transformed into a series of news bulletins about Martians invading the state of New Jersey. "No one involved with *War of the Worlds* expected to deceive any listeners, because they all found the story too silly and improbable to ever be taken seriously," *Smithsonian Magazine* reports. "The Mercury's desperate attempts to make the show seem halfway believable succeeded, almost by accident, far beyond even their wildest expectations."[287] Were there really stampedes as people fled their homes in the wake of the fake news of a Martian invasion? Did some listeners really threaten to shoot Welles on sight the next day because of the panic and damage his prank broadcast had caused? Perhaps, though there is some evidence as well that "the newspaper industry sensationalized the panic to prove to advertisers, and regulators, that radio management was irresponsible and not to be trusted."[288] A case of fake news begetting fake news, if ever there was one. Is it any wonder that there is a contingent of folks who believe the Apollo moon landing was a hoax and Elvis faked his own death?

While we're on the subject, however, we cannot discount the idea that we may believe—and *want* to believe—fake news, particularly if it confirms our previously held biases and desires. I'd certainly like to believe Elvis was still around and making music to a ripe old age, and I'm sure I'm not alone in that wish. But what of people who believed, at the height of the 2020–2021 pandemic, and absent any credible research by medical professionals, that hydroxychloroquine was a cure for COVID-19? What of those who, even months after Joe Biden has been sworn in as president, and every trustworthy vote audit in the land has confirmed his legitimacy, continue to assert that Donald Trump won the 2020 election? What of

---

[287] A Brad Schwartz, "The Infamous 'War of the Worlds" Broadcast Was a Magnificent Fluke," *Smithsonian* Magazine (May 6, 2015), https://www.smithsonianmag.com/history/infamous-war-worlds-radio-broadcast-was-magnificent-fluke-180955180/.

[288] Mark Memmott, "75 Years Ago, 'War Of The Worlds' Started a Panic. Or Did It?" *NPR* (October 30, 2013), https://www.npr.org/sections/thetwo-way/2013/10/30/241797346/75-years-ago-war-of-the-worlds-started-a-panic-or-did-it.

those folks who will still swear up and down that it rained on the largest crowd ever in human history assembled for Trump's inauguration, in spite of contemporaneous weather reports and photographic evidence? "'It's inexpensive—and in fact cheaper—to produce lies rather than truth, which creates conditions for a lot of false information in the marketplace,' says Harvard Law Professor Noah Feldman, an expert in constitutional law and free speech who was instrumental in the development of Facebook's Oversight Board, 'We still collectively have a tendency to believe things we hear that we probably shouldn't, especially when they seem to confirm prior beliefs we hold.'"[289]

Again, while you're pondering man-bats, Martian invasions, and the efficacy of hydroxychloroquine on the coronavirus, keep in mind that it is the ability to disseminate the misinformation and disinformation quickly and on a global scale that is new. In an October 2020 article for *Scientific American*,[290] Joan Donovan, adjunct professor at Harvard Kennedy School and research director of the Shorenstein Center on the Media, told the story of the Yes Men, a group of artist-activists who, around 1999, perpetrated a series of hoaxes by way of the internet as a strategy to advance their social and political agenda. They were able to spoof the truth with such dexterity and success that they were, at times, offered speaking engagements as the entities they were spoofing, such as the National Rifle Association, Shell, the *New York Times*, and the World Trade Organization. However, as Donovan points out, the Yes Men might really have been too good at what they did. "Just a couple of decades later technology companies have created a media ecosystem that allows governments, political operatives, marketers and other interested parties to routinely expose internet users to dangerous misinformation and dupe them into amplifying it. There is mounting evidence of foreign

---

[289] Elaine McArdle, "Oh, What a Tangled Web We Weave," *Harvard Law Today* (July 7, 2021), https://today.law.harvard.edu/feature/oh-what-a-tangled-web-we-weave/.

[290] Joan Donovan, "Trolling for Truth on Social Media," *Scientific American* (October 12, 2020), https://www.scientificamerican.com/article/trolling-for-truth-on-social-media/.

operatives, partisan pundits, white supremacists, violent misogynists, grifters and scammers using impersonation on social media as a way to make money, gain status and direct media attention."[291]

Social media platforms "play a major role in the diffusion and amplification of misinformation and disinformation narratives," according to scientists at the Harvard Kennedy School.[292] Let's take a quick tour of how, according to Donovan, they accomplish the work of spreading the messages of those foreign operatives, partisan pundits, and all the rest.

First, content on the internet is *repetitive*; that is, it will often be posted by more than one party to assure the widest audience possible for the information or, of course, the disinformation. Next, it is *redundant*, meaning that it isn't usually posted on just one platform but spread out over the entire social media ecosystem, from Facebook to YouTube to Twitter to Pinterest and more. Third, *responsiveness* is at the core of social media interactions—meaning that we are compelled to respond to posts with which we both agree and disagree, and to act as translators of the content for friends and followers in our own circles. In either case, this assures that the post or ad gets more attention from the platform's algorithm. And that brings us to our fourth and final "r," which is *reinforcement*. Once triggered by a user's engagement, the algorithm will keep sending the same message or similar messages to the user's newsfeed.

There is, however, one final ingredient that can contribute to the amplification of any social media message, and that is the use of automated social media accounts, known more commonly as "bots."

"At the very basic level a bot is a social media account which is preprogrammed to do stuff without people pressing the button," says Ben Nimmo, the information defense fellow for the Atlantic Council's Digital Forensic Research Lab. He goes on to say, however, "There are plenty of good bots out there, like there are bots that share poetry. What we focus

---

[291] Ibid.

[292] "Tackling Misinformation: What Researchers Could Do with Social Media Data," *Harvard Misinformation Review* (December 9, 2020), https://misinforeview.hks.harvard.edu/article/tackling-misinformation-what-researchers-could-do-with-social-media-data/.

on particularly is malicious bots which are bots used to distort conversations online."[293] Botnets, which are networks of bots that can range in size from only a few hundred to a collection of millions, can amplify a single Facebook post by, well, a million-fold, if the network is large enough.

While these are the causes behind the dissemination of misinformation and disinformation *today*, I really can't leave behind this discussion of how fake news is spread without touching on at least one of the factors that stands to make the problem much worse in the future: the expanding capabilities of artificial intelligence (AI). Henry Kissinger, former U.S. Secretary of State and National Security Advisor under the administrations of Richard Nixon and Gerald Ford and co-author of *The Age of AI: And Our Human Future*, has said that "AI could be as consequential as the advent of nuclear weapons—but even less predictable";[294] we'd be remiss not to pay attention.

In May 2021, the Center for Security and Emerging Technology (CSET) issued a report about GPT-3, a powerful, cutting-edge AI system designed by a company called OpenAI. While to date disinformation has been a fundamentally human undertaking,[295] the GPT-3 system may just upend the foundations. Already, "among other achievements, it has drafted an op-ed that was commissioned by *The Guardian*, written news stories that a majority of readers thought were written by humans, and devised new internet memes."[296] CSET warns that a system this formidable could help "disinformation actors substantially reduce the work necessary

---

[293] Bob Abeshouse, "Troll Factories, Bots and Fake News: Inside the Wild West of Social Media," *Aljazeera* (February 8, 2018), https://www.aljazeera.com/features/2018/2/8/troll-factories-bots-and-fake-news-inside-the-wild-west-of-social-media.

[294] Meredith Wolf Schizer, "Henry Kissinger Says AI Is 'as Consequential' but 'Less Predictable' Than Nuclear Weapons," *Newsweek* (November 2, 2021), https://www.newsweek.com/2021/11/12/henry-kissinger-says-ai-consequential-less-predictable-nuclear-weapons-1644508.html.

[295] Ben Buchanan, Andrew Lohn, Micah Musser and Katerina Sedova, "Truth, Lies, and Automation," *CSET* (May 2021), https://cset.georgetown.edu/publication/truth-lies-and-automation/.

[296] Ibid.

to write disinformation while expanding its reach and potentially also its effectiveness."[297] In other words, at some point in the near future we may experience automated fake news generated and distributed at scale.

While in the present we might have a more tenuous attachment to objective truth than generations before us, losing our grip entirely on a common community narrative could bring to our institutions, and the society that upholds them, the destruction that we have so far managed to avoid.

It is the "layering of information," according to Donovan, that makes you believe it. Or, in the words of Marshall Shepherd, "According to a 2015 study in the *Journal of Experimental Psychology: General*, the *illusory truth effect* is the notion that repeated statements are perceived to be more truthful than new statements. This effect is clearly something that marketing professionals, cult leaders, and politicians understand. In other words, you say something enough times, and people start to believe it."[298]

Now that you have a solid understanding of the sort of mind games marketers, politicians, and other interest groups play with us, and the strategies behind their deployment, let's take a look at some of the real-world, real-time consequences of our steady and pervasive diet of illusory truth.

---

[297] Ibid.

[298] Marshall Shepherd, "Misinformation Doesn't Make It True, But Does Make It More Likely to BE Believed," *Forbes* (August 17, 2020), https://www.forbes.com/sites/marshallshepherd/2020/08/17/why-repeating-false-science-information-doesnt-make-it-true/?.

Chapter 4

The Algorithm and the Damage Done

Long ago, in a galaxy that, for our youngest generations, can likely seem farther away than Tatooine, the world experienced its first truly global counterculture. A "counterculture" is one wherein the norms, behaviors, and values of its adherents differ in substantial ways from the mainstream mores of the day. History is certainly peppered with examples.

The Levellers were a political movement in Britain that emerged during the chaotic time of the English Civil War (1642–1651). They believed in equality, human suffrage, and religious tolerance, and appealed to the disaffected in their society through the distribution of pamphlets and public speeches. They advocated for a plan that would have thoroughly democratized England, and their ideas gained traction even among factions of the army. Their plan, however, also threatened the supremacy of Parliament, and the movement, at the outset aligned with Oliver Cromwell, was crushed by him in 1649 in what became known as the Banbury Mutiny.

We could also cite examples such as Bohemianism, which began to take root in Paris in the 1800s and for which, by the 1850s, New York's Greenwich Village had become a prominent hub; the Non-conformists, who emerged in the period between the World Wars and tried to find a "third alternative" between socialism and capitalism; or the Beats, who in the 1950s first popularized the exploration of Eastern religions, the rejection of consumerism, and, importantly, the honest and explicit portrayals of the human condition in poetry and fiction. But counterculture is by no means, of course, an exclusive feature of the West. Aleksandr Solzhenitsyn stands as one of the most prominent figures to emerge from Russia's anti-communist, pro-human-rights movement, a cohort of loosely affiliated groups and individuals called the Dissidents who emerged in 1953, following the death of Josef Stalin. Mahatma Gandhi is the personification of the Indian Independence Movement, of which he became the leader in 1917.

What we're talking about right now, however, is the Sixties counterculture: the hippies. The hippie movement arose in America out of a collective desire to bring an end to the war in Vietnam, but it rapidly evolved to encompass a myriad of other issues, and its ideology made its way around the globe. From promotion of racial desegregation to the fight for women's rights, from an emphasis on a back-to-nature way of living and "natural" foods to efforts to ban censorship, from fighting widespread poverty to fighting environmental pollution, from the hair and clothing styles that became emblematic of its adherents to the drug culture and psychedelic rock that became associated with it, it embodied almost everything that was "anti-establishment." And it spread from the United States to the cities of Western Europe, like Paris, London, West Berlin, Rome, and Amsterdam. It spread to the Czech Republic, where long-haired young people were arrested and forced to cut their hair. It spread all along the "hippie trail," where travelers primarily from North America, Western Europe, Australia, and Japan toured Turkey, Iran, Nepal, and beyond.

From each of these counterculture movements we can find at least a seed from which the world, as we know it, has grown. But no movement has had such a direct—we could even say glaring—impact on our contemporary lives as the counterculture of the hippies of the 1960s. While their compatriots were exploring Turkey and Nepal, there was a cohort of anti-establishment types back home in Silicon Valley who were unsatisfied that "mainframes (whose cost and complexity kept them in the hands of corporate-sponsored priesthoods), or business-oriented personal computers (such as the IBM PC),"[299] kept the wonders of computing out of the hands of the masses. Indeed, one of Apple's first advertising slogans, back in 1984, was "the computer for the rest of us." "As inventor Jim Warren put it, the personal computer 'had its genetic coding in the 1960s'…antiestablishment, antiwar, profreedom, antidiscipline attitudes.'"[300] Those young visionaries might not have known, or been able to articulate it in quite this way, but what they were dreaming up was personal computers, cell phones, and a commercialized internet.

"Every innovation starts with an act of insubordination," said futurist and scientist Walter de Brouwer.[301] In the case of the wonders of our contemporary technological world, the insubordination came from pure ideals and laudable motives, and even if they hadn't, could we in good conscience fault yesterday's innovators when their innovations have so greatly improved our everyday lives? Among those conveniences we take for granted today, what would we be willing to give up in order not to look like hypocrites if we were to place blame on the advances they've made? Email? Your children's school portals? Your cell phones? How about your favorite app—the pedometer that tracks how many steps you've taken in the course of a day, or the service that plays all your favorite tunes while you're driving, or the one that allows you to order-in lunch with just a few clicks? How about the home security system that keeps your house and your family safe?

Most of us can truthfully say that we both understand the good intentions behind our present technological conveniences and are grateful for them. Moreover, we can agree that we're inspired that those ideals still flourish in the newest generations of innovators—those young

---

[299] "Computing and the Counterculture," *Stanford.edu* (November 6, 2021), https://web.stanford.edu/dept/SUL/sites/mac/counter.html.

[300] Ibid.

[301] Adrienne Burke, "How The '60s Counterculture Is Still Driving The Tech Revolution," *Forbes* (November 21, 2013), https://www.forbes.com/sites/techonomy/2013/11/21/how-the-60s-counterculture-is-still-driving-the-tech-revolution/?sh=2cdd41cb6dd3.

visionaries and inventors who are right now dreaming up the next big, game-changing advances that we'll find ourselves equally unwilling to give up in the future.

However, none of our gratitude for the past's innovators or the future's inspirations precludes us from being practical about the present. The novelist Paulo Coelho wrote, in the global phenomenon that is his book *The Alchemist*, that "Every blessing ignored becomes a curse." The wisest course for us, in the here and now, is to turn our attention to assessing what is good about our present technological lives, and what is less good, and how we will mitigate the damage as we become ever more dependent on ever more advanced technologies.

To that end—so that we don't ignore the blessings that come with being fully wired in—let's take a look at four specific areas where technological advances have made, or are on the cusp of making, sweeping changes in the way we live. Let's then talk about what we want to do to repair or mitigate the damage, while we still have the opportunity to conceptualize the most desirable, and the least intrusive, ways in which tech can impact our lives.

When No News Is Bad News

Picture this scene: a woman pouring morning coffee or perhaps an evening cocktail, while a man is seated at a breakfast table or in a den (the precursor to the man cave) reading a newspaper. This scene is a cliché so prevalent in old movies, TV shows, and advertisements that even the youngest readers will be able to conjure a mental image, albeit likely in black and white. The scene, however, was not merely a cliché, but shorthand. Depending on how the actors or models were posed, we could tell if we were getting a glimpse into a happy home with a contented husband and wife, or one that was maybe not so happy, with the man pointedly ignoring his spouse and using the newspaper as a sort of a shield against her. The classic television show *I Love Lucy* once did a sharp role-reversal take on the ubiquity of husbands catching up on the morning's news before they went out to work in the world[302]—house*wives*, of course, having a lesser need for such information—which stands as sort of a funny, poignant elegy to the once-pervasive presence of a morning and/or an evening newspaper in the American home. Either way, a happy-if-traditional home or an unhappy one, newspapers were simply a feature of everyday life, evoking a certain stability, not only in the home but in the larger world. They were an essential element in the American Dream scene.

The operative word in that last sentence is *were*. Since 2004, more than two thousand American newspapers have gone out of business,[303] and a whole lot of other papers are having a hard time staying afloat. Now, while profit margins for newspapers that enjoy national distribution, such as the *New York Times*[304] and the *Washington Post*,[305] have ebbed and flowed over the last decades—sometimes dipping precipitously—the defunct newspapers were, for the most part, local media, part of the landscape of small cities and towns across the nation. What has caused this mass demise? Primarily, the shift to digital content.

Print journalism has faced competition consistently for a hundred years, as technologies advanced and brought radios and then televisions into people's homes. We're far from the days when we relied on broadsheets—and I'm speaking here of those large, old-time informational sheets of paper printed on one side only—as our primary news source. Yet, newspapers survived, and even thrived, through the decades when said radios and TVs did offer alternative sources for news. Paper boys, and sometimes girls, kept their jobs through the 1950s, '60s, '70s, and beyond. The competition the papers face in the internet era, however, is materially different, so let's break it down.

---

[302] "I Love Lucy Breakfast English," *YouTube*, https://www.youtube.com/watch?v=nbxUo6OuXTg.

[303] "The Decline Of Local News," *NPR* (August 3, 2020), https://www.npr.org/2020/07/30/897134561/the-decline-of-local-news.

[304] "New York Times Profit Margin 2006–2021," *Macrotrends*, https://www.macrotrends.net/stocks/charts/NYT/new-york-times/profit-margins.

[305] "Post Holdings Profit Margin 2010–2021," *Macrotrends*, https://www.macrotrends.net/stocks/charts/POST/post-holdings/profit-margins.

First, there is the issue of the translation of content from one medium to a wholly different one. In some ways, this mirrors the shift reporters and copywriters had to make as they tailored their stories to appeal to readers, then listeners, then watchers. The detail one could fit into a newspaper story would almost universally not fit into a five-minute radio news report read reliably at the top of the hour; the visuals that drew viewers to TV news shows were, by definition, unavailable to their print and radio competitors. Indeed, it was not until November 22, 1963, following the assassination of President John F. Kennedy and four days of solid, uninterrupted coverage, that television became an "essential"[306] part of the contemporary media mix—and a critical part of the coverage of the Vietnam war, antiwar protesters, inner-city riots, and the funerals of two more beloved, legendary leaders: the Reverend Dr. Martin Luther King, and Senator and presidential candidate Robert F. Kennedy.

Yet, the competition for readers and listeners and viewers didn't preclude the continuing success of each form of media. Each medium had its uses, advantages, and place in the mix. When the internet entered the picture, however, life got more complicated for print and air-time salespeople, including those who sold spots for the emergent cable news networks. This was not because the platforms offered by Big Tech used a different advertising model; as we talked about earlier, the model for media advertising sales was the brainchild of Ben Franklin and it has, so far, truly stood the test of time. It was because users consumed news differently on the internet.

"The *[New York] Times* was always interested in finding new ways of conveying news in the fastest possible way to the greatest number of people," said David W. Dunlap, historian for the *Times*. "That is: We were platform-agnostic from the moment [Publisher, Adolph S.] Ochs

---

[306] Stephen Battaglio, "War, riots and assassinations: 1968's TV news paved the way for today's endless, opinion-heavy coverage," *Los Angeles Times* (March 23, 2018), https://www.latimes.com/business/hollywood/la-fi-ct-1968-tv-news-20180323-htmlstory.html.

and [Managing Editor, Carr V.] Van Anda took over, if not earlier."[307] What he was referring to was the unique, and, frankly, ingenious ways in which the *Times* had conveyed information to its readership in the past. On the eve of Election Day 1904, as an example, the *Times* announced it would be flashing searchlights from the top of its new building on 43rd Street in Times Square to indicate the election results: if a steady beam was shown to the east, that would mean Democrat Alton B. Parker had won the election, and if the beam was shown to the west, that would indicate incumbent Theodore Roosevelt had been re-elected.

However, in the face of the challenge of the internet, remaining "platform agnostic"—which refers to software products that run equally well across a number of platforms—was not going to be enough to keep readers' attention. They were growing used to the bells and whistles that internet content offered to them—real-time updates to the big stories of the day, colorful photographs to illustrate each story, videos on demand, the ease of scrolling without getting newsprint all over one's fingers, all of it for much less than the price of a subscription to a print newspaper, and often even free. In the fast-paced arena of internet news, simply shifting what you would print anyway to a page on your website didn't give readers enough of a reason to stick around. Indeed, newspapers—the old-fashioned print sort—were at what you could call a "platform-agnostic disadvantage," in that they had no framework for nor ability to gather data about their readers and target their audience anywhere near as efficiently as could their digital counterparts.

Add to these facts that, on the internet, a reader wasn't limited to one or two papers but could access a growing number of websites to deliver the news. For example, Bloomberg.com went online in 1993; *The Economist's* site and whitehouse.gov debuted in 1994; Salon.com went live in 1995; Slate.com followed in 1996. And, on the web, a reader wasn't limited to only U.S. news, of course: the U.K.'s *Daily Telegraph* went online in 1994, as did the *Irish Times*. It was around this same time

---

[307] Mark Shimabukuro, "Who Won the Election? Years Ago, the Sky Had the Answer," *New York Times* (November 6, 2020), https://www.nytimes.com/2020/11/06/insider/election-light-signals.html.

that smaller-circulation newspapers, such as the *News & Observer* in Raleigh, N.C., began to show up with online editions, as did American newspapers devoted to amplifying ethnic voices, including Cuban and Asian newspapers, and historic Black ones, such as Baltimore's *The Afro*, Philadelphia's *Philadelphia Tribune*, and New York's *Amsterdam News*. With so many sources to choose among, a reader was no longer confined by the four corners of a printed newspaper, but found a world of news, ideas, and opinions opened to him from the comfort of his own home, with only a few clicks on his computer screen.

Along with new, digital methods of delivery, and an entire world of online newspapers now at hand, came whole new modes of reporting, too. For reporters, that new mode is known as *open-source journalism*, a practice that takes advantage of the myriad "open source," meaning "free"—that is, freely available to the public at large—sources available now to reporters by way of the internet, including some tools that were once reserved for use by the military and other governmental agencies, such as satellite imagery. Using these open sources allows reporters access to a wide range of materials and information that was not available—or not available with the simple click of a mouse—twenty or even fifteen years ago. Using data garnered from disparate sources enables reporters to make connections that might not have been readily apparent without the advantage of these digital tools, and it helps them tell broader, deeper stories as a result.

Hand-in-hand with open-source journalism came *citizen journalism*, or the reporting of news events from members of the public using digital cameras, smartphones, and the internet to broadcast their messages and/ or get news items noticed by larger, more mainstream media. Darnella Frazier, the teenager who recorded George Floyd's murder on her cell phone, is an exemplary example of a citizen journalist. News that might have gone unnoticed—cops that might have gotten away with abusive behavior—finds the light of day when nearly every person walking the street has only to reach into a pocket or a purse to easily document every encounter along the way.

o o o

There were other ramifications, from each individual having a whole world of open-sourced news outlets to choose among—not to mention of news sources now able to find a substantial audience, not from only one small geolocation but from the whole world. One is that the news became more niche. Don't misunderstand the history here: some newspapers have always been partisan to one extent or another. Indeed, most newspapers in the eighteenth and nineteenth centuries were "tied to a political party and their content on political matters heavily favored that party."[308] Remember that one factor in Ben Franklin's decision to buy the *Pennsylvania Gazette* was that he wanted a forum for his own political opinions. The sort of fierce polarization within the United States that took root in the last twenty or thirty years can be attributed, at least in part, to the plethora of exclusively partisan news and opinion sources of both left and right.

The polarization of the populace is such an important matter that we'll take it up later in this chapter, in a whole section of its own. For the moment, let's recognize that producing the news—paying, for example, the investigative journalists and other writers, as well as the copy editors and graphic artists and other technicians needed to make it slick—is not an inexpensive proposition. In a world with such a surfeit of news sources, where each source is vying for the same pool of advertising dollars, often resorting to "click bait"[309] to draw in and keep readers, perhaps this kind of partisanship was inevitable as traditional, general interest newspapers died and sources that remained struggled to remain viable. Cutbacks and layoffs necessitated by the coronavirus pandemic and associated economic downturn of 2020–2021 are also playing a part in the struggle for survival the newspaper business is now enduring. But, no matter the reason,

---

[308] Shigeo Hirano and James M Snyder, Jr. "Measuring the Partisan Behavior of U.S. Newspapers, 1880 to 1980," Columbia University, https://www8.gsb. columbia.edu/media/sites/media/files/newspaper_partisanship.pdf.

[309] According to Merriam-Webster, "something (such as a headline) designed to make readers want to click on a hyperlink especially when the link leads to content of dubious value or interest.

the fact remains that, as of 2019, one in five U.S. newspapers have gone out of business;[310] in 2020 alone, more than 8,200 newspaper employees lost their jobs.[311]

At one time, a reliable source of a newspaper's revenue was the classified ads section. Potential employers would pay to take out an ad one-column-width wide and a few inches long, to post that they were looking for a secretary or a babysitter, a forklift operator or floor manager or a sales clerk. Landlords would pay to post that they had an apartment or maybe a house to rent. Someone who was cleaning out his garage would pay to post that he had a lawnmower, in good condition, to sell or trade. But Craig's List debuted in 1996, and people could post their classified ads for free, so even that source of newspaper revenue was reduced or went away entirely.

As small newspapers around the country were going out of business, the populations that had once depended on them for local, state, as well as national news, shrugged their shoulders. A 2011 Pew poll, for example, found that "a large majority of Americans, 69%, believe the death of their local newspaper would have *no* impact (39%) or only a *minor* impact (30%) on their ability to get local information."[312] Further, "Younger adults, age 18–29, were especially unconcerned. Fully 75% say their ability to get local information would not be affected in a major way by the absence of their local paper. The same was true of heavier technology users: 74% of home broadband users say losing their paper would have no impact or only a minor impact on their ability to get local information."[313]

[310] Lara Takenaga, "More Than 1 in 5 U.S. Papers Has Closed. This Is the Result," *New York Times* (December 21, 2019), https://www.nytimes.com/2019/12/21/reader-center/local-news-deserts.html.

[311] Timothy Karr, "Making Big Tech Pay News Outlets Won't Help Save Independent Journalism," *Truthout* (March 18, 2021), https://truthout.org/articles/making-big-tech-pay-news-outlets-wont-help-save-independent-journalism/.

[312] "Perceptions of the Importance of Local Newspapers," Pew Research Center (September 26, 2011), https://www.pewresearch.org/internet/2011/09/26/part-3-the-role-of-newspapers/.

[313] Ibid.

On the other hand, the same Pew study reported that among all adults, newspapers were still considered "the most relied upon source or tied for most relied upon,"[314] with respondents conceding that "dependence on newspapers for so many local topics sets it apart from all other sources of local news."[315] This is heartening to hear. To explain why, let's take a short detour into our civic lives.

Most of us understand that a primary responsibility of civic life is voting. Unfortunately, too many of us interpret that to mean *voting in a presidential election*. While casting a ballot for president every four years is certainly important, so is every other state and local election along the way. It is these elections—and those who win them—that build the foundation upon which our democracy stands. Local school board members determine the budgets and curriculums for local schools, and therefore play an integral part in determining the quality of education the children in the community will receive. It is the city or county sheriff who is charged with keeping the local peace and enforcing local laws. It is the city or town mayor and her or his council who propose and enact laws and ordinances over a wide range of local issues—city budgets and taxes, water and waste services, and the recruitment of new business, to cite just a few. If not for our local newspapers, where do we find information on the incidence of local crime, the rate of local taxes, the activities of local government, the availability of local jobs and housing and social services? Moreover, the fulfillment of our civic responsibilities requires more than only voting; it requires us to be active participants in our society. Except for our local news sources, where are we to come by information about our community, arts, and other cultural events happening around us? We're not going to be able to find such useful information in the *New York Times* or the *Washington Post*, unless we happen to live in New York or D.C. It isn't going to be written about in either *Salon* or *The Daily Wire*, or carried on CNN, FOX, or MSNBC.

Democracy starts at home, with even the smallest election, town hall, and school board meeting. It begins and flourishes with each of

---

[314] Ibid.
[315] Ibid.

us knowing about the events going on right around us, in our neighborhoods, school districts, towns, and cities. Having fewer trusted *local* news outlets leaves us less informed. That makes us less effective and engaged citizens. And that means when it comes to the big things, the big elections, we are collectively less equipped to participate, and likely less enthused about it, too. It's just human nature to feel not particularly connected to those things we don't understand. Or care about. And that could be how democracy dies.

Racial Animosity

It isn't any secret that America has a profound racial injustice problem. Read the headlines, take a look around you—the evidence is everywhere that a 400-year-old problem still exists and that, as a nation, we're still trying to come to grips with our understanding of it and its origins. Take, as an example, the debate raging furiously at the time of this writing around the necessary work of teaching critical race theory.[316]

It's not within the scope of this book, however, to take on such a large and ongoing crisis and fix the injustice. The politics, the history, or even the urgent, contemporary movement to do something concrete to find definitive solutions are beyond these pages. What is within our focus is how racial injustice continues to manifest as it pertains to Big Tech. This may strike some readers, at first glance, as if I'm carving out a less-than-significant sliver of the story of American race relations. Really, how can inanimate, electrical objects like computers have anything to do with the human social construct of race? But, as I will show you, the discussion around even one of the issues at the intersection of race and tech is emblematic, in microcosm, of the insidious ways racism continues to manifest itself in our lives.

When I began working on this book, one of the first bits of research I did was to read *Algorithms of Oppression* by Safiya Umoja Noble.

---

[316] "What Is Critical Race Theory, and Why Is Everyone Talking About It?" Columbia University (July 1, 2021), https://news.columbia.edu/news/what-critical-race-theory-and-why-everyone-talking-about-it-0.

Dr. Noble is an Associate Professor at the University of California, Los Angeles (UCLA), in the Department of Information Studies, and Co-Founder and Co-Director of the UCLA Center for Critical Internet Inquiry (C2i2). In her 2018 book she describes the way information is allotted to Google's search engine users as "technological redlining."[317]

Redlining is the practice of systematically denying goods or services, the word itself rooted in the color-coded system put in place in the 1930s by the Home Owners' Loan Corporation (HOLC). The HOLC was part of the New Deal policies and programs designed to pull the country out of the Great Depression. The HOLC's part was to help people refinance their mortgages and avoid foreclosure, and their color-coded system was supposed to help banks decide where to extend mortgages and mortgage assistance. Green indicated very desirable neighborhoods that were good investments; blue, acceptable ones; yellow, those in decline; and a red line drawn around a neighborhood meant that it was "hazardous" or "risky." Property values within the red lines were low and loans shouldn't be extended to potential homeowners who lived or wanted to buy within the red outline. Given what we know of history, it won't surprise you to know that the green neighborhoods were white neighborhoods and the red ones were Black. "We must remember that, because everyone was poor during the Great Depression, these maps did not reflect economic status. In fact, upscale black neighborhoods like LaVilla and Sugar Hill in Jacksonville, Florida, home to Duke Ellington, Ella Fitzgerald, and Zora Neale Hurston, were deemed 'too risky,' by the HOLC."[318]

Noble makes her point that Google's search engine offers its services with the same sort of "bias-free neutrality" as the HOLC employed when choosing which homeowners would benefit from its programs. She holds

---

[317] Safiya Umoja Noble, "Algorithms of Oppression: How Search Engines Reinforce Racism," https://www.amazon.com/Algorithms-Oppression-Search-Engines-Reinforce-dp-1479849944/dp/1479849944/ref=mt_other?_encoding=UTF8&me=&qid=1621800144.

[318] Michael Harriot, "Redlining: The Origin Story of Institutional Racism," *The Root* (April 25, 2019), https://www.theroot.com/redlining-the-origin-story-of-institutional-racism-1834308539.

out as her first example the results she received when she typed the phrase "Black girls" into Google's search bar. She had been looking for activities she might enjoy on a day out with her pre-teen niece and the niece's friends. But what appeared on her screen was a display of sexualized Black girls and women, a spectacle she could not allow the young girls in her charge that day to see.[319] *Search neutrality* is a concept that means search engines should insert no editorial prejudgments into the process, but re-sults should be offered to the user in a manner that is comprehensive and impartial. The results of Noble's search for "black girls" was evidence that Google's algorithm was hardly impartial; it was returning results based on some of the topmost stereotypes about Black women—that is, that they are sassy, loud, angry, and overtly sexual.

I finished reading this portion of Noble's book and sat down at my computer to conduct my own experiment. With some trepidation, I typed "black girls" into my search bar. You can imagine my relief when the first two results of my search were two films, both titled *Black Girl*. The 1966 film, directed by Ousmane Sembène, was hailed as the beginning of African cinema by the respected film critic Jonathan Rosenbaum, and the 1972 film was directed by Ossie Davis, an inductee to the NAACP Image Awards Hall of Fame. I allowed myself a moment to wonder that, perhaps, major changes had been made between the time when Noble was writing her book in 2018, and I was writing this one in spring 2021. Certainly, people had been working to correct the problem of racial bias in algorithms for some time. It was in 2017 that the New York City Council passed the first-in-the-nation bill "to help ensure the computer codes that governments use to make decisions are serving justice rather than inequality."[320]

But the moment passed. I knew full well that tech companies had not completely overhauled their algorithms, bending the arc of the moral

---

[319] "Search Engine Breakdown," *PBS* (April 14, 2021), https://www.pbs.org/wgbh/nova/video/search-engine-breakdown/.

[320] Rashida Richardson, "New York City Takes on Algorithmic Discrimination," *NYCLU* (December 12, 2017), https://www.nyclu.org/en/news/new-york-city-takes-algorithmic-discrimination.

universe, in such a short time period. In any case, companies are noto-riously secretive—proprietary—about the codes that make their algo-rithms work, so discovering the nuances of why I was spared soft porn in my "black girls" search, while Noble was not, would likely be an exercise in futility.

What I can say for sure is that the biases remain, even if you or I can't detect them in our search results. Take the case of Optum, a leading health services company whose algorithms provide a "guide to [health] care decision-making for millions of people."[321] This was a widely-used program to determine which patients needed additional care and which did not, but a study reported in the journal *Science* found that "Black patients assigned the same level of risk by the algorithm are sicker than white patients. The authors estimated that this racial bias reduces the number of Black patients identified for extra care by more than half."[322]

By more than half. Why? Because their algorithm used "health costs as a proxy for health needs. Less money is spent on Black patients who have the same level of need, and the algorithm thus falsely concludes that Black patients are healthier than equally sick white patients."[323]

Eliminating the racial discrimination from this particular algorithm, then, would seem to be an easy fix: stop using health costs as the proxy for health needs. How, though, do you cure an algorithm like Twitter's that "prefers younger, slimmer, faces with lighter skin"?[324] Bogdan Kulynych,

---

[321] Carolyn Y. Johnson, "Racial bias in a medical algorithm favors white pa-tients over sicker black patients," *Washington Post* (October 24, 2019), https://www.washingtonpost.com/health/2019/10/24/racial-bias-medical-algorithm-favors-white-patients-over-sicker-black-patients.

[322] Ziad Obermeyer, Brian Powers, Christine Vogeli and Sendhil Mullainathan, "Dissecting Racial Bias in an Algorithm Used to Manage the Health of Populations," *Science* (October 25, 2019), https://science.sciencemag.org/content/366/6464/447.full.

[323] Ibid.

[324] Alex Hern, "Student Proves Twitter Algorithm 'Bias' Toward Lighter, Slimmer, Younger Faces," *The Guardian* (August 10, 2021), https://www.theguard-ian.com/technology/2021/aug/10/twitters-image-cropping-algorithm-pre-fers-younger-slimmer-faces-with-lighter-skin-analysis.

a graduate student at Switzerland's EFPL university demonstrated that the company's cropping algorithm, which was designed to focus on image previews of "the most interesting parts" of photos, preferred younger, slimmer, lighter faces over older, wider, darker ones. Remarking on his discovery, Kulynych said, "Algorithmic harms are not only 'bugs'. Crucially, a lot of harmful tech is harmful not because of accidents, unintended mistakes, but rather by design. This comes from maximization of engagement and, in general, profit externalizing the costs to others. As an example, amplifying gentrification, driving down wages, spreading clickbait and misinformation are not necessarily due to 'biased' algorithms."[325]

The problem, then, as we see, is not encompassed by one particular algorithm or another, and the root of it goes much deeper than one quick fix. As Eric Griffith, writing for *PC Magazine*, put it, "Racial bias in tech products tracks back to an industry that continues to believe the 'default' internet user is white, male, and cisgender."[326]

And that "default" tracks back to the people who work in tech. According to a 2020 Internet Health Report conducted by Mozilla, a five-year study finds that the diversity among tech companies is not a model for any Diversity and Inclusion (D&I) department. In 2014, for example, Apple was made up of 54 percent white employees, 23 percent Asian employees, 7 percent Latinx, and 6 percent Black. By 2019, those percentages hadn't changed much at all: 49 percent white, 35 percent Asian, and 8 percent Latinx, with Black employment remaining at 6 percent.[327] The percentages are in the same ballpark for Facebook, Microsoft, and Google, except that these three companies have an even lesser percentage of Latinx and Blacks in their work forces.[328]

---

[325] Ibid.

[326] Eric Griffith, "Algorithms Still Have a Bias Problem, and Big Tech Isn't Doing Enough to Fix It," *PC Magazine* (January 28, 2021), https://www.pcmag.com/news/algorithms-still-have-a-bias-problem-and-big-tech-isnt-doing-enough-to.

[327] Mozilla website, "A Healthier Internet Is Possible," https://2020.internethealthreport.org.

[328] Ibid.

Remember what algorithms are: a set of instructions, or a *recipe* a computer follows to deal with a data set, or *a list of ingredients* to come to a designated outcome or *dish*. You give ten different chefs the same set of ingredients, and you're going to get ten different dishes, all of them seasoned by the chef's culinary history, training, experience, and personal favorite flavors. But all these algorithmic recipes are written by the people who work in the tech industry, the vast majority of whom are white, cisgender men. This by no means is to suggest a wholesale dismissal of every white, cisgender man who works in tech as a racist. I am suggesting, however, that, as humans, they bring their own lived experience to bear on the jobs they do. Of course, this is not, in and of itself, a bad thing—we all do it, and often this provides value to our employers who benefit from our perspectives. When a company lacks diversity on the scale that tech currently lacks it, however, that company has little chance of accurately reflecting the lived experiences of its users and customers, and serving them in the most relevant way.

Let me end this section, however, on a positive note, by taking notice of one small but sensible step forward. As Axios reports, "Historically, the chemical processes behind film development used light skin as its chemical baseline—basically optimizing for whiteness, a legacy that continues today, says Snapchat engineer Bertrand Saint-Preux. 'The camera is, in fact, racist,' Saint-Preux said."[329] In April 2020, in response, the good folks at Snapchat have "launched an initiative to redesign its core camera technology to make it better able to capture a wide range of skin tones."[330] Over at Google, they are focused on issues around a color scale known as Fitzpatrick Skin Type (FST). FST has been used by dermatologists for fifty years, and tech companies picked it up for use in measuring how certain products such as facial recognition systems work equally well over a range of skin tones. The problem? FST has four categories for "white" skin and only one each for "black" and "brown" skin, completely ignoring

---

[329] Ina Fried, "1 big thing: How Snapchat is trying to make the camera more inclusive," AXIOS (April 29, 2021), https://www.axios.com/newsletters/axios-login-682b5969-fbfa-412a-851e-49adf866e4b1.html.

[330] Ibid.

diversity among people of color. In June of 2021, Google said that it had been quietly pursuing a better system: "We are working on alternative, more inclusive, measures that could be useful in the development of our products, and will collaborate with scientific and medical experts, as well as groups working with communities of color."[331]

These are two attempts to address the on-going cultural problem of whiteness as the racial default—meaning this isn't a problem exclusive to tech. And, on the scale of earth-shattering progress in terms of racial justice, it is, as I said, a small step, along the same lines as Johnson and Johnson's rollout of a line of bandages in a diverse color palette, which it did in June 2020 after years of being urged to do so by advocates.[332] But if we can stay focused on the big picture—on the larger issues around racial justice that we still have to solve—we can celebrate these small but welcome gestures.

How Smart Is Smart-Home Living?

In 1962, an animated television show call *The Jetsons* premiered on ABC. The show, created by Hanna-Barbera, was in many ways just like their hit show *The Flintstones*; but where the Flintstones were a typical sitcom family living in the prehistoric era, the Jetsons lived in 2062—exactly one hundred years from the date the show made its TV debut. And it depicted life in all its futuristic glory with absolutely stunning accuracy.

The Jetsons enjoyed videoconferencing and cell phones. They watched television shows on their wrist watches and got home in driverless cars. They had a robotic maid to do the vacuuming. They even had a family pet, Astro the Space Mutt, whom they walked on a treadmill. In all

---

[331] Paresh Dave, "EXCLUSIVE Google Searches for New Measure of Skin Tones to Curb Bias in Products," *Reuters* (June 18, 2021), https://www.reuters.com/business/sustainable-business/exclusive-google-searches-new-measure-skin-tones-curb-bias-products-2021-06-18/.

[332] "Band-Aid Adding New Brown and Black Skin Tones," *CBS Boston* (June 12, 2020), https://boston.cbslocal.com/2020/06/12/band-aid-new-colors-skin-tone-black-brown-johnson-and-johnson/.

honesty, the latter had been around as a concept since Ancient Rome—that is, a "tread wheel" was used by the ancient Romans as a replacement for cranes when they had to lift a very heavy object; but treadmills didn't take their ubiquitous place in gyms and even the home until the 1970s. The Jetsons lived an almost entirely automated, hands-free life, but without any of the downsides we are obligated to think about, as the functionality of automation and AI improve, and as we gradually but steadily remove ourselves from the sort of typical daily labor that running a home now necessitates.

For a first and obvious example, the Jetsons were remarkably fit for people who traveled from room to room in their home on moving walkways, and had to do little more than press a button to make dinner or buy a new dress. While a whole lot of us have grown used to checking our daily steps on either a Fitbit or an app on our phones (constantly striving for that 10,000-steps-a-day goal in lives that could become quite deskbound and sedentary if we didn't police ourselves), the only Jetson who seemed to get any exercise was George, the father of the family, whenever he joined Astro on the treadmill. I'll just note here that Fitbit, one of those connected devices we can wear on our wrists like a bit of jewelry, became part of Google in 2021; and in the announcement that they'd become part of the "Google family," Fitbit went of its way to assure "that users' health and wellness data won't be used for Google ads and this data will be kept separate from other Google ad data."[333] The fact remains, however, that the data will still be collected, and the myriad other uses to which it can be put remain undefined.

Still, when you get past your concerns about the Jetsons' cardiovascular health, their smart home is enviable. It's equipped with a plethora of electronic gadgets that service every need, small or large, with the press of a button, as well as lighting and climate control that appear to function on intuition alone. It's a standard this cartoon family seems to have set for real-life developers of smart-home technology: your home senses when

---

[333] James Park, "FitBit Joins Google," *FitBit,* https://blog.fitbit.com/2021-up-date/?utm_source=go_mar&utm_medium=blog&utm_country_code=en_US&utm_campaign=web.

you walk into a room and the lights will turn on; when you walk up a set of stairs, aisle lights follow your footsteps and illuminate your way; when you arrive home from a day of work, your house is so attuned to your routine that your garage door opens of its own volition to welcome you home; and the heat or air conditioning unit have anticipated your return and already begun the process of coming to temperature, so you're as cozy or as cool as the season demands as soon as you walk through your front door. In the cartoon—as in various video tours of actual smart homes created in large part by developers of the technology in use in the home—the tech works flawlessly, all its functions coordinated with seamless precision so your at-home hours are, essentially, thoughtless. You don't have to concern yourself with cleanliness because the robot vacuum cleaner has taken care of sweeping up. Even food—the way we grow it, prepare it, and consume it—is reconsidered for smart-home purposes. That consideration spans 3D food printers that use ground and/or dehydrated, and often additive-riddled, raw materials to produce edibles, to laboratory-grown "meat" that scientists believe might be an environmentally sustainable way to adequately feed the world's growing population,[334] to innovations that allow you to grow healthier consumables right on your own windowsill.[335]

However, what you will have to consider in a smart-home scenario is cost. While the cartoon family presents the ownership of all of its remarkable technology merely as elements of the typical, middle-class lifestyle of the future, creating a smart home in reality can be expensive and out of reach for large swaths of the population. Gearbrain is a website created to explain how the Internet of Things works, and to help an average consumer navigate the marketplace in order to efficiently design her own custom smart home. They estimate a median cost for smartening up the average three-bedroom, two-bathroom house to be around

---

[334] Jonah Mandel, "Israelis Taste the Future with Lab-Grown Chicken 'Food Revolution,'" *Yahoo News* (June 21, 2021), https://news.yahoo.com/israelis-taste-future-lab-grown-153024189.html.

[335] Chloe Rutzerveld, "3D Printed Food: The Future of Healthy Eating," YouTube, https://www.youtube.com/watch?v=hw321SwC6kA.

$15,000—and that's just for the basics such as lighting, a thermostat, a video doorbell, motorized blinds, and a hub to connect all the devices and functions. Adding a smart refrigerator, smart exercise bike, smart bathroom scale, media streaming sticks for every television, and/or an upgraded lighting or media center will drive the costs higher.[336]

Perhaps more to the point: what about those of us who may not want or be able to invest, at least all at once, in a complete smart-home setup and who choose to improve our homes' technology in a more piecemeal fashion—perhaps using less expensive components that we install ourselves? That was what tech writer Adam Clark Estes decided to do. He used components made by the Wink smart-home system, which is marketed and sold at a friendlier price point—thus, as Estes put it, taking smart homes out of the exclusive hands of the "tech-hungry rich."[337] How did that work out for him? "Pretty much as soon as my Wink hub arrived, the problems started, and it didn't take long for that excitement to turn into frustration. Without going into the mind-numbing details of each individual point of frustration, let me sum up the experience very bluntly: Not a single Wink product worked as advertised right out of the box."[338]

Say that you have invested in the hardware, software, hours of time, and frustration it takes to make your home smart. How do you *keep* it smart? Your next real concern is installing a back-up system, so you have an alternative plan in place in case there is an electrical outage or your power grid fails. Owning a lighting system that automatically turns on, and even adjusts itself to whatever mood you're usually in at any given time of the day, loses its value if the lights won't turn on at all. The number of gadgets and systems you will need to acquire in order to create, maintain, and improve your smart home might well turn out to be as

---

[336] Alistair Charlton, " How Much Does It cost to Turn the Average Home into a Smart Home?" *GearBrain* (February 8, 2021), https://www.gearbrain.com/average-smart-home-build-cost-2589554626.html.

[337] Adam Clark Estes, "Why Is My Smart Home So F***ing Dumb?" *Gizmodo* (February 12, 2015), https://gizmodo.com/why-is-my-smart-home-so-fucking-dumb-1684949715.

[338] Ibid.

infinite as the imagination of the engineer who designs these products and systems. And the body of knowledge the average consumer will need to obtain and assimilate to keep his home functional takes us right back to a discussion we had earlier and to British historian Ruth Goodman, who holds that homemaking in the future will require an ever-expanding, practical understanding of tech.

o o o

Concerns about the smart-home lifestyle go well beyond reliability of the systems we put into place and the expense of acquiring them. Neil Jacobstein, chairman of the Artificial Intelligence and Robotics Track at Singularity University, warns that, over the next ten to twenty years, a full "47% of U.S. jobs are vulnerable to automation and, around the world, the numbers are even higher."[339] That's a nearly overwhelming disruption in our direct and immediate future, though if we view it from an historical perspective, it can seem slightly less daunting. Throughout the ages not just types of jobs, but whole categories of them have gone by the wayside. Can you name, off the top of your head, anyone who runs a livery stable or manufactures buggy whips? What is the market for milkmen or milliners these days? When was the last time you met an elevator, linotype, or switchboard operator? Have you even ever heard of a gandy dancer—the slang term for the workers who laid and maintained railroad tracks before this work was done by machines—or a powder monkey, a member of the crew of naval warships whose job was to ferry gunpowder from the powder magazine in the ship's hold to its artillery pieces?

As technologies change and come into common use, the job market changes as well. There may not be a livery stable in the heart of most towns any longer, but there are gas stations, automotive sales lots, car parts stores, and mechanics' garages all over the place—not to mention the jobs in automobile factories and oil refineries that exist because very few of us ride our horse into town to shop at the general store anymore.

---

[339] "Neil Jacobstein: AI, Automation, and the Future of Technology," *YouTube* (February 2, 2017), https://www.youtube.com/watch?v=HTBF1skaDUg.

The problem with the way we're dealing with the enormous job skills shift that's headed our way is that we aren't properly preparing people for the jobs to come. "Safety nets and training systems built up over decades to protect workers are struggling to keep up with the 'megatrends' changing the nature of work," Bloomberg has reported, adding the unsettling fact that six out of ten workers currently lack basic IT skills.[340]

An article by the Pew Research Center[341] outlines five distinct directions in which job training may change to meet the challenge of the evolving employment landscape:

1. The current educational system will evolve, migrating in great part to online learning, in an employment atmosphere where workers must learn continually in order to maintain skills necessary to match rapidly advancing technologies. Universities and colleges will diversify their offerings to accommodate new knowledge and skills required for their students to be employable after graduation.

2. Soft skills such as emotional intelligence, curiosity, creativity, and critical thinking will increase in value, as will experiential learning through apprenticeship programs.

3. The credentialing system will evolve. Though college degrees will still be important for some jobs, proof of competency in a real-world environment may overtake a glittering resumé as a primary job qualification.

4. Naysayers insist that job training will not meet twenty-first-century needs within the next ten to fifteen years. Politicians lack the will necessary to fight for the money that will be necessary

---

[340] William Horobin, "Automation Could Wipe Out Almost Half the Jobs in 20 Years," *Bloomberg* (April 25, 2019), https://www.bloomberg.com/news/articles/2019-04-25/the-future-of-work-could-bring-more-inequality-social-tensions.

[341] Lee Rainie and Janna Anderson, "The Future of Jobs and Jobs Training," *Pew Research Center* (May 3, 2017), https://www.pewresearch.org/internet/2017/05/03/the-future-of-jobs-and-jobs-training/.

to provide appropriate jobs training and, anyway, people are incapable or uninterested in continuing education.

5. Advances in technology will leave millions unemployed, and capitalism itself may collapse as the work force is fundamentally unable to meet the needs of the market.

I realize that list was a rapid descent from a highly optimistic scenario of hope and competence to one in which the world as we know it, in a matter of a few decades, spirals into one that would be unrecognizable to us today. In at least one real way, unfortunately, we've already started that descent: unskilled or less skilled workers are already being replaced by automation in many industries. This job loss to automation is, in turn, fueling the wage and earnings inequality that has risen sharply in the United States, as well as in other industrialized countries, over the last four decades. A report issued by the National Bureau of Economic Research tells us that "The real earnings of men without a high school degree are now 15% lower than they were in 1980."[342] The report continues, underscoring the need for appropriate education in the new digital era, if we are to have a work force both prepared to do the jobs of the future and fairly compensated for the work they perform: "Workers who are not displaced from the tasks in which they have a comparative advantage, such as those with a postgraduate degree or women with a college degree, enjoyed real wage gains, while those, including low-education men, who used to specialize in tasks and industries undergoing rapid automation, experienced stagnant or declining real wages."[343]

This is not to say that there aren't whole categories of jobs for which, in the past, higher learning was a priority that might also be on the block. For example, the *Miami Herald* is experimenting with a new "employee"—the *Miami Herald* Bot that, working from templates prepared

---

[342] Daron Acemoglu and Pascual Restrepo, "Tasks, Automation, and the Rise in US Wage Inequality," *National Bureau of Economic Research* (June 2021), https://www.nber.org/system/files/working_papers/w28920/w28920.pdf.

[343] Ibid.

by actual, human journalists, is covering the real estate beat for the newspaper. Sometimes the bot gets the stories right, and other times "its articles do not make much sense at all"[344] but, as an editor's note that appears at the Bot's stories puts it, "We are experimenting with this and other ways of providing more useful content to our readers and subscribers."[345] And the *Miami Herald* isn't alone in this experiment: "Over the past several years, journalists at Forbes, the Associated Press, Bloomberg, and the Guardian have developed bots to cover everything from breaking news to sports to politics."[346]

In short, we are already feeling, in very real ways, the pain from having neglected to prepare our work force for the jobs that need to be done today and into the future. The pain is only going to get worse the longer we put off tackling the problem.

It doesn't have to *stay* painful, however. That's why I want to give the last word on this to Vivek Wadhwa, a tech entrepreneur and Distinguished Fellow at the Labor and Worklife Program at Harvard Law School. What if we had a work schedule more on the order of our cartoon father, George Jetson, whose full-time job was toiling away at Spacely's Sprockets, turning the Referential Universal Digital Indexer (R.U.D.I.) on and off for a grueling nine hours a week. What if, Wadhwa asks us to imagine, we completely reimagined what it meant to hold a full-time job? What if humans really do take the reins, retrain ourselves, and create automation that serves our needs so efficiently and humanely that we need work only ten hours a week?[347] What then? What would you do with a life so different from the ones we think of today as typical?

---

[344] Alex Deluca, "Meet the Miami Herald's Newest Writer: A Robot!", Miami New Times (November 4, 2021), https://www.miaminewtimes.com/news/miami-herald-robot-writes-real-estate-stories-13219683.

[345] Ibid.

[346] Ibid.

[347] Vivek Wadhwa, "The Future of Work Is No Work," *YouTube* (August 27, 2018), https://www.youtube.com/watch?v=bCh6rqL1HWQ.

o o o

The question above—what *would* you do with a life in which you could readily support yourself and your family by working only ten hours a week?—leads nicely to our final consideration, which is: will smart-home living make us stupid? In other words, what will making our homes smart take away from us? This question isn't about only all the dreamy leisure pursuits we'd take up if we had more time to do so, but about those skills and abilities that might well atrophy if we turn over now-human responsibilities to automation.

For instance, if we end up creating our meals on 3D printers, what becomes of family traditions like Sunday dinner and the simple, satisfying pleasure of cooking real food for the people we love? What becomes of the world's cuisines, or recipes passed down from generation to generation, if we abdicate in the kitchen? It wouldn't be only the ability to create sustenance we'd be giving up, but a whole creative outlet—the thinking process involved in following a recipe or the innovation required to invent one. Kitchen skills are already in short supply among younger generations, with 60 percent of Millennials confessing they don't know how to make a salad dressing, and 25 percent expressing doubt they'd be able to make an edible cake using a box mix.[348]

And what of those moving walkways? Right now, most of us encounter them most commonly when we're rushing through an airport; and when you're dragging a fifty-pound suitcase behind you, they can certainly seem like a blessing. But what happens if they become ubiquitous? What if technology really does offer us paths to even more sedentary lives than we're faced with now? A sedentary lifestyle leads to real harms, and the ones most quantifiable are those associated with our health: obesity, type 2 diabetes, cardiovascular disease, some cancers, and early death.

---

[348] Maura Judkis, "Do millennials really not know how to cook? With technology, they don't really have to." *Washington Post* (April 12, 2018), https://www.washingtonpost.com/news/voraciously/wp/2018/04/12/do-millennials-really-not-know-how-cook-with-technology-they-dont-really-have-to/.

Let's start with obesity. Prolonged periods of inactivity—when we're sitting at a desk behind our computers, or lying in bed scrolling through social media on our phones, or slumped on the sofa in front of the TV—can impair the body's metabolism. That can lead to the impairment of the body's ability to regulate our blood pressure, our blood-sugar levels, and the way fat is broken down. Rising levels of obesity have been a problem for humans ever since we stopped working in the fields to harvest our own food to eat, or walking to meet up with friends instead of hopping in the car, or even hitching up our horses to ride into town. We are reaching a crisis point: in a mere three-year period, America went from just six states with a 35 percent adult obesity prevalence in 2017, to nine states in 2018, to twelve states in 2019.[349] Broken down by gender, we see that in the 1960s, somewhere around 16 percent of American women were considered medically obese; by the 2010s, that percentage had risen to over 30 percent.[350] Men fare even worse in the same comparison. In the 1960s, just over 10 percent of men were considered medically obese; by the 2010s, that percentage had risen to nearly 40 percent.[351] The "shoe leather express" is defined by the Urban Dictionary as a sarcastic term for someone having to walk somewhere. It's a mode of transport with which we might want to become familiar once again, before we lose our health and our mobility to tech.

o o o

Speaking of modes of transportation, we can't very well end this section about smarter-living-through-tech without talking about cars.

---

[349] "Obesity Worsens Outcomes from COVID-19," *Centers for Disease Control and Prevention* (September 17, 2020), https://www.cdc.gov/media/releases/2020/s0917-adult-obesity-increasing.html.

[350] Cheryl D. Fryar, MSPH; Margaret D. Carroll, MSPH and Cynthia L. Ogden, PhD, "Prevalence of Overweight, Obesity, and Extreme Obesity Among Adults: United States, Trends 1960–1962 Through 2009–2010," *Centers for Disease Control and Prevention*, https://www.cdc.gov/nchs/data/hestat/obesity_adult_09_10/obesity_adult_09_10.htm.

[351] Ibid.

Electric vehicles (EVs) have been on the radar for decades, and manufacturers are lately committing to some rapidly approaching dates by which they will be selling mostly or, in some cases, only electric vehicles. "General Motors says it will make only electric vehicles by 2035, Ford says all vehicles sold in Europe will be electric by 2030 and VW says 70% of its sales will be electric by 2030."[352] Before we know it, electric cars will become standard and, close on the heels of that, driverless cars will be the next big thing. Dozens of tech companies are working feverishly to get their own models on the roads, and we've been inching toward a mass deployment of this technology for a while now. So we should take a moment to define what a "driverless car" actually is.

Most new cars these days already come equipped with some level of automation. Think self-parking capabilities, cruise control, and GPS navigation. A truly driverless car—or self-driving car, or autonomous vehicle, which have become fairly interchangeable terms—is one that can move from one location to another without any intervention from a human driver. It comes equipped with various technologies that continually monitor its surroundings, such as radar, lidar, laser lights, and GPS, to name just a few. Highly trained AI can detect and react to factors such as constantly changing road conditions and unanticipated obstacles in the way of its progress.

Perhaps the best known among the various driverless cars is Tesla. While Elon Musk has long been the public face of the company, it was actually founded in 2003 by Martin Eberhard and Marc Tarpenning, with the mission of creating affordable, mass-market electric vehicles. Musk didn't become involved until 2004, investing $30 million in the company and becoming chairman of its board of directors. And it wasn't until 2013 that the company entered the race to produce a driverless car.

Tesla has recently endured a series of crashes of its cars that were fitted with its partially automated driving system called Autopilot. Looking at only the crashes that occurred immediately around the time of this writing, we have one in April of 2021 in Texas, in which the car crashed

---

[352] Justin Rowlatt, "Why Electric Cars Will Take Over Sooner Than You Think," *BBC News* (June 1), https://www.bbc.com/news/business-57253947.

into a tree, killing both people who were inside;[353] another fatal crash in Fontana, California, on May 5, 2021, in which the driver was killed and a pedestrian seriously injured when the car hit an overturned semi on the freeway;[354] and a third that occurred on May 19, 2021, when a 2015 Tesla Model S car hit a parked police vehicle. No one was hurt in the May 19th crash, but "the incident caused 'significant damage' to the patrol car."[355]

There is also Waymo, which began its life as a Google project but was later sliced off to become a separate company for Google's parent, Alphabet. Waymo started out in the driverless-car niche with a huge advantage—Google's considerable experience with AI. What it didn't have was car manufacturing experience. So Waymo teamed up with automakers like Fiat Chrysler Automobiles (FCA), Volvo, Renault, Nissan, and Magna (which isn't itself an automobile brand but makes vehicles for Jaguar and Chrysler). There is Apple's driverless car, being developed under the name Project Titan; Uber's, which operates under the name Uber Advanced Technologies Group; and Microsoft's project, called AirSim, which involves developing internet-connected vehicles within its portfolio. In the commercial trucking space, Embark Trucks[356] already has self-driving trucks on the road. It has achieved many "firsts" in this fast-advancing technology, including trucks that have successfully driven across the country, operated in rain and foggy conditions, and navigated

---

[353] "Tesla Autopilot Crashes and Causes," *AutoPilot Review* (Updated September 2021), https://www.autopilotreview.com/tesla-autopilot-accidents-causes/#:~:text=Most%20common%20causes%20for%20Tesla%20Autopilot%20crashes%20fall,false-positive%20sudden%20braking%20%28which%20could%20cause%20more%20accidents%29.

[354] "Feds Investigating Another Fatal Tesla Crash; California Man Arrested in Back Seat of Driverless Car *MarketWatch* (May 12, 2021), https://www.marketwatch.com/story/feds-investigating-another-fatal-tesla-crash-california-man-arrested-in-back-seat-of-driverless-car-01620864532.

[355] Tesla in Autopilot Mode Crashes into Parked Police Car," *ABC News* (May 19, 2021), https://www.msn.com/en-us/autos/news/tesla-in-autopilot-mode-crashes-into-parked-police-car/ar-BB1gS3ky?ocid=uxbndlbing.

[356] Embark website, https://embarktrucks.com/story.html.

between transfer hubs. Embark is poised to make commercial traffic safer and potentially reduce the costs of shipping by up to fifty percent.[357]

Driving—getting one's driver's license as a teenager—was once considered a rite of passage. If the inventors—visionaries?—of these autonomous vehicles have their way, and can perfect the technologies necessary to put a truly self-driving car safely on the roads, that milestone for so many sixteen-year-olds may become a thing of the past. At the other end, no one may ever again have to have that awkward conversation with an elderly parent whose fading eyesight or reaction times make it necessary for them to surrender their driver's license.

The convenience that could come with a self-driving car is abundantly self-evident, as is the idea that these driverless cars will make the roads much safer for travel. Driverless car boosters cite studies that say human error causes from 90 to 99 percent of traffic accidents[358]—though clearly *all* accidents caused by drunk driving are caused by the human error of getting behind the wheel of a moving vehicle while under the influence of an intoxicating substance. There were 10,142 deaths caused by humans driving drunk in 2019,[359] the last year for which we have complete statistics. Imagine if a similar death toll could be prevented in the future by driverless cars. Here's the most salient fact about the safety of driverless cars: when one of these cars is involved in an accident, the on-board AI learns from that accident how to prevent it or lessen its impact—and is able to apply this knowledge on its journeys forward from that point. Even more exciting is that—unlike human drivers, who might or might not learn as individuals from one accident and then apply that lesson to one individual car—when AI learns, its advanced knowledge can be applied to *every other driverless car on the road*, through the simple act of periodically updating the software on board other driverless vehicles, too.

---

[357] In the interests of disclosure, my husband is involved in Embark.

[358] Bryant Walker Smith, "Human Error as a Cause of Vehicle Crashes," *Stanford Law School, The Center for Internet and Society* (December 18, 2013), http://cyberlaw.stanford.edu/blog/2013/12/human-error-cause-vehicle-crashes.

[359] United States Department of Transportation, National Highway Traffic Safety Administration, https://www.nhtsa.gov/risky-driving/drunk-driving.

While this technology is still in development, what is really most important is that we take the opportunity to consider the larger matters concerning living in a world made "smarter" by technology. The moment's questions are: How will all these artificially intelligent machines and gadgets not only benefit but potentially harm us? How will they limit or curtail our skills and thinking processes? Will they make us dumb? Will, for example, the skill of driving a car become as recreational as that of saddling and riding a horse in twenty years' time?

○ ○ ○

In chapter seven, we'll talk more, and in more depth, about what the future might hold for us, including the shape warfare will take if fought with ever more technologically advanced weapons. For now, let's focus on what might happen if the United States suffers a cyberattack, and if the target of that attack is our power grid—a recognized vulnerability.

There are three interconnected power grids in the United States: the Eastern Grid, the Western Grid, and the Texas or ERGOT Grid. The nation got a taste of the crisis that would ensue if our grid collapsed when just one portion of it, the Texas Grid, suffered a major electricity-generation failure in February of 2021. The failure resulted in shortages of water and food; in the dead of a winter storm, over 4.5 million homes were left without power and heat; and approximately 150 people lost their lives, though there are reports that the death toll was five times the number that was officially released.[360] Take a minute to imagine the chaos and panic that would result in the event of a nationwide outage. Some experts tell us that, in a scenario such as this, we would be thrust back in time to a pre-electrical era—that is, to the lifestyle of a year pre-1882, when the country's first electrical system, the Pearl Street Station in lower Manhattan, began operation. We would become reliant on wood or coal

---

[360] Jamie Ross, "Texas Winter Storm Might Have Killed Five Times Number of People State Admits," *Daily Beast* (May 27, 2021), https://www.thedailybeast.com/texas-winter-storm-might-have-killed-five-times-number-of-people-state-admits?ref=home.

to heat our homes and cook our dinners, on ice blocks to keep our food from spoiling, and on candles to light our way in the dark.

Other experts disagree. They tell us that reverting to the 1880s would be an optimistic outcome in the case of a major attack on our grid system. They predict we would revert to a much earlier time, such as the 1400s, because absent technology and other modern services, very few people any longer have the skills necessary to live lives as evolved as our Victorian ancestors. Who among us knows how to efficiently chop wood, for example, or even has access to a forest where he might do this harvesting? How many of us have working wood-burning fireplaces where we can safely ignite this fuel? Who not only knows how to hunt and butcher, but has the tools on hand to do these things? Who will harvest ice from a riverbank, cut it into blocks, and cushion it with hay so it will not melt completely before it can be delivered to us? Who and what horse is going to be the iceman cometh to our door? Do you know how to make candles, and do you have either enough animal fat or beeswax at hand to make enough to keep your home bright each and every night after the sun goes down? No?

Welcome back to the Middle Ages.

This might seem like a dramatic example, but it highlights the point I want to make, which is how ill-equipped we have become as a civilization to face daily life without technology at our fingertips. Herbert Marcuse (1898–1979) was one of the most prominent and best-known philosophers and social theorists in the world in his time. In his 1969 work, "An Essay on Liberation," he wrote, "The so-called consumer society and the politics of corporate capitalism have created a second nature of man which ties him libidinally and aggressively to the commodity form. The need for possessing, consuming, handling and constantly renewing the gadgets, devices, instruments, engines, offered to and imposed upon the people, for using these wares even at the danger of one's own destruction, has become a 'biological' need." He was, in this assessment as in so many others he made in the course of his work, far, far ahead of his time.

We are poised at the threshold of yet another quantum leap forward in our technological lives. I strongly recommend that we define for ourselves, for our service providers, and for our government what our needs for technology really are, as well as the terms under which we're willing to take advantage of the advances Big Tech is offering to us. Where is the balance between drudgery and agency? The trade-off between convenience and creativity? Where is the line in the sand beyond which privacy becomes a hard priority?

In the future, your house may well have a record—not to mention an actual video recording—of every move you make. Every phone call placed from your smartphone, every television show streamed on your smart TV, every morsel of food in your smart fridge. And your autonomous car may well have a record of every mile you've traveled, and every stop you've made along the way to your destination. Now, today, well before technology is priced at a point that the vast majority of us can afford truly functional smart homes and truly safe driverless cars, is the time to negotiate how much we want our homes—and therefore the companies that power them—to know about us, who owns this information, and how it can be used.

We may, in many ways, be living in a live-action, real-life version of *The Jetsons*. But Big Tech is living in the Wild West, where rules are few, weak, and ineffectively enforced, and where there aren't really any meaningful boundaries at all, because the sheriffs who are supposed to be keeping order can't find the keys to the jail.

Now is the time for us, as thinking consumers, to decide where the boundaries are going to fall.

Busted: The Cambridge Analytica Scandal

One particular case study can underscore the need to create and enforce boundaries. The case study I'm talking about is a relatively recent one, and more or less familiar to all of us who follow national politics. It illuminates quite clearly how loose the boundaries are around tech right now, and what havoc is let loose in our lives because we continue to allow tech to function in this slack, gray area.

Before we dive into the actual study, let's begin by taking a few steps back and revisiting the idea of "clickbait." That is online content whose main purpose is to attract attention with some sort of sensational news or claim, and so compel a user to click on a link. But let's be honest: as a concept, attracting eyes with sensational news or claims certainly didn't start with the internet. From splashy headlines in newspapers to bulletins announcing "breaking news!" on radio and television, sensationalism has long been used to acquire readers, listeners, and viewers, in order to drive up subscriber numbers and garner Nielsen ratings. From sales pitches (think newsboys, or "newsies" as they were known, hawking a supplement to the daily paper by shouting, "Extra! Extra! Read all about it!" in the days before radio brought fresh news more quickly to the populace) to grand publicity stunts (think Anheuser-Busch sending their iconic Clydesdale horses to the White House to deliver a case of Budweiser to President Roosevelt on the day Prohibition was repealed), we consumers have given advertisers every reason to believe spectacle is the most effective way to get our attention.

Are they wrong? Well, which of the following newspaper accounts of a local fire would you rather read?

First account: *A minor fire broke out in the kitchen of XYZ restaurant last evening during the dinner rush. The fire started due to a grease build-up in the hoods, but kitchen workers were able to easily put it out with baking soda, though a cautious hostess evacuated diners for their safety until the fire chief was able to make a full examination and clear the restaurant to continue dinner service.*

Second account: *A grease fire erupted in the kitchen of XYZ restaurant last evening while the restaurant was packed with diners. Customers rushed the exits as a panicked hostess tried desperately to herd them into the parking lot. Kitchen staff worked heroically to douse the flames with ordinary baking soda and it was this quick thinking that saved the lives of the diners, and the building itself from ruin. Miraculously, after a full investigation, the fire chief cleared the restaurant to continue dinner service.*

Same fire, same diners, same restaurant, and same outcome. But based on their coverage, which reporter is going to draw in more readers with her story? And which reporter is the editor more likely to send out to cover the next breaking news event?

What has changed from the days when salacious news came to us via newsprint is the delivery system. The historical fact is that disinformation and conspiracy theories have been around for as long as humankind has had an ear for gossip and an affinity for spicy information. From the rumors that Shakespeare didn't write his own plays and that Queen Elizabeth I was really a man, to more current ones—such as the moon landing was faked, 9/11 was an inside job, and the Sandy Hook shooting was the work of child actors—a certain mindset has always been drawn to the fantastical and macabre. "American politics has often been an arena for angry minds," the historian Richard Hofstadter wrote in 1964, in an attempt to codify this state of mind he deemed the "paranoid style."[361] The difference is that today a story—honest and true, vigorously and even maliciously embellished, or made up completely from whole cloth—can go viral in the first thirty seconds after it's posted on the internet.

There are concrete reasons behind any viral story. Let's revisit a few key points.

First, dispel the notion that the reason for virality is that any particular news story or social media post contains inherent value, and we are simply passive consumers of content that happens to appear in our feeds. The stories we see in our feeds are fed to us by algorithms that determine the value of the story based on several factors, all them rooted in how much attention the story has already received. So, how does a story get this coveted attention?

First, *bots*. We've already talked about bots in chapter three, but their influence is so great that it's worth taking another look at exactly what they are, and the forms in which they come. A bot is an automated software application programmed to perform certain specific tasks, and a bot network is a number of bots—a few hundred or a few million—programmed to perform the same certain specific task, and so amplify

---

[361] Richard Hofstadter, "The Paranoid Style in American Politics," *Harper's Magazine* (November 1964), https://harpers.org/archive/1964/11/the-paranoid-style-in-american-politics/.

the message or task of the bot network. There are many types of bots. There are *search-engine bots* that index information so it can be efficiently presented to users who are searching for content. There are *service bots* that act as customer service to assist users in, for example, finding information or making a purchase. Among these are *chatbots*, which are programmed to simulate human conversation, and *web crawlers*, which are sometimes referred to as Googlebots, that scan content on web pages in order to index it for consumer use. Then there are *social bots*, which are programmed to interact with social media users. While some social bots are benevolent or neutral, most of them are malicious, programmed to *scrape* content—that is, to remove all content from a website, nearly always without the site owner's permission, and often to repurpose it for nefarious reasons; to perpetrate a *credential stuffing attack*, which is the act of taking information from one data breach and using it to attempt to infiltrate other, unrelated sites or services (for instance, taking account login information gleaned from hacking a bank's web service and using those passwords and other information to attempt to make purchases on Amazon); or to *spread* disinformation and other spam content.

Kate Starbird, a University of Washington professor who studies how conspiracy theories propagate online, describes a botnet as an "'alternative media ecosystem'—a complex network of bots, individuals and domains that birth, orchestrate and promote conspiracy theories that undermine online readers' trust in information."[362] Botnets, deployed by disinformation peddlers, are one of the main ways conspiracy theories and other false information make their way into our newsfeeds and even into the discourse of more general, mainstream news sources—even if they appear there only in order that they can be debunked.

A story can also gain traction because of the "likes", "retweets", and "comments" it gets from actual humans on a social media site. A site's algorithm takes notice of any attention a post is receiving, and the more

---

[362] Rebecca Heilweil, "How Internet Conspiracy Theories Go Viral—And Get People To Believe Them, Too," *Forbes* (June 25, 2017), https://www.forbes.com/sites/rebeccaheilweil1/2017/06/25/how-internet-conspiracy-theories-go-viral-and-get-people-to-believe-them-too/.

"likes" or other interactions it gets, the more it is fed into your feeds and those of your social media "friends". Your "likes" are weighted, depending on which "like button" option you choose to respond to any one post: "like", "love", "care", "laugh", "wow", "sad", or "angry".[363] Simply clicking that you "like" a post doesn't do much for it, in terms of influencing the algorithm; but if you express a stronger emotion—"love" or "anger", for example—that pricks up the metaphorical ears of the algorithm and lets it know that this post may be worth more in terms of its emotional value. In this way, middle-of-the-road perspectives are filtered out, to be replaced by highly emotional or even inflammatory ones.

According to Scott Galloway—New York University professor, author of *The Four*, and tech entrepreneur—Facebook, for example, "knows it's being used by troll farms to spread lies that divide us, dial up the distrust and outrage and get us marching to extremes. 'The true believers, whether from the left or the right, click on the bait,' says Galloway. 'The posts that get the most clicks are confrontational and angry. And those clicks drive up a post's hit rate.'"[364] The shorthand here is that when a particularly juicy story breaks, whether it's true or false, it's how we—along with a potential army of bots—respond to the juice that's going to give it legs.

Nowhere are the results of this sort of online manipulation more abundantly clear than in the case of Cambridge Analytica—a now-defunct, London-based election consultancy best known for its strong ties to the 2016 presidential campaign of Donald J. Trump, as well as Britain's Brexit vote in March of the same year. Thanks to the courage of two whistleblowers—Brittany Kaiser and Christopher Wylie, both former

---

[363] Jenna Amatulli, "Facebook Prioritizes What Makes You 'Sad' Or 'Angry' Over What You 'Like'" *Huffington Post* (March 3, 2017), https://www.huffpost.com/entry/facebook-weighs-your-reactions-more-than-your-likes_n_58b-7044ce4b019d36d0fe13c.

[364] Jay Robb, "The Four Horsemen of God, Love, Sex and Consumption," *The Hamilton Spectator* (October 23, 2017), https://www.thespec.com/business/2017/10/23/jay-robb-the-four-horsemen-of-god-love-sex-and-consumption.html.

Cambridge Analytica employees—more than 100,000 documents relating to the work of the company in sixty-eight countries were laid out to reveal the extensive, global network this "operation used to manipulate voters on 'an industrial scale.'"[365] How did Cambridge Analytica manage this industrial-scale manipulation? Primarily by misappropriating the personal information contained in over 87 million Facebook profiles.[366]

Let's be blunt about this point: the misappropriation of our personal information, and its deployment to manipulate us, is done for one reason and one reason only: profit. That profit can be, and often is, money. For example, did you know that hate speech has a definite monetary value? "While quantifying the value of the market for disinformation is difficult, it is clearly a major source of income for conspiracy theorists, hate groups, and disinformation media moguls. According to a July 2020 report by the Center for Countering Digital Hate (CCDH), the online audience for anti-vaccination content may be generating nearly $1 billion in revenue for social media companies and reaching an audience as large as 58 million people. A joint report published in October 2021 by the Institute for Strategic Dialogue (ISD) and the Global Disinformation Index (GDI) found that online retail is one of the most common funding sources for hate groups.[367] The conspiracy theorist Alex Jones has made a fortune selling supplements and other merchandise on his own website, as well as e-commerce platforms. As of October 2021, Infowars supplements were still available for sale on Amazon."[368]

---

[365] Carole Cadwalladr, "Fresh Cambridge Analytica Leak 'Shows Global Manipulation Is Out of Control,'" *The Observer* (January 4, 2020), https://www.theguardian.com/uk-news/2020/jan/04/cambridge-analytica-data-leak-global-election-manipulation.

[366] Ibid.

[367] "The Business of Hate, Bankrolling Bigotry in Germany and the Online Funding of Hate Groups, Institute for Strategic Dialogue and Global Disinformation Index (2020), https://disinformationindex.org/wp-content/uploads/2021/09/GDI-ISD_The-Business-of-Hate.pdf.

[368] Patrick Jones, "The Conspiracy and Disinformation Challenge on E-Commerce Platforms," *Brookings* (June 10, 2021), https://www.brookings.edu/techstream/the-conspiracy-and-disinformation-challenge-on-e-commerce-platforms/.

Or, the profit can also be about power—or a combination of both money and power—but the motive is always profit, done in order to control our thought processes, our votes, our pocketbooks, and, ultimately, *us*. What Cambridge Analytica did—and what others continue to do—is to weaponize our personal information against us, for the benefit of those who seek to attain or maintain a profit at our expense.

Many of us were left in shock and awe as the extent of just this one, massive incident of digital manipulation and its consequences were uncovered. There was the personal affront of finding out that our personal autonomy and that of our friends, families, and fellow citizens—our power to come to our own conclusions and make our own decisions based on those conclusions—had been degraded. There was real, palpable fear as we realized how this manipulation had polarized and destabilized our nation, compromising the United States's long standing as the world's essential democracy. And there was anger, frustration, and confusion as we wondered if there was anything we could do to stop bad actors from doing it again—if we could keep it from happening over and over into the future, continuing to weaken us, personally, as well as to debase and even destroy the institutions the world has relied upon for centuries.

danah boyd[369]—a Partner Researcher at Microsoft Research, the founder and president of Data & Society, and a visiting professor at New York University—summarizes the problem we must solve in order to rebuild our social cohesion and allay fears that our society may not be strong enough to withstand ongoing attacks from those who seek to manipulate us. "All too often, technology is designed naively, imagining all of the good but not building safeguards to prevent the bad. The problem is that technology mirrors and magnifies the good, bad AND ugly in everyday life. And right now, we do not have the safeguards, security or policies in place to prevent manipulators from doing significant

---

[369] danah boyd styles her name in lowercase.

harm with the technologies designed to connect people and help spread information."[370]

What we can do to install those safeguards, security features, and policies will be the focus of the next chapters of this book.

---

[370] Janna Anderson and Lee Rainie, "Many Tech Experts Say Digital Disruption Will Hurt Democracy," *Pew Research Center* (February 21, 2020), https://www.pewresearch.org/internet/2020/02/21/many-tech-experts-say-digital-disruption-will-hurt-democracy/.

Chapter 5

Your 'Net Worth

Let's play make-believe for a moment. Let's say that you and I are neighbors. We each have a house with a big suburban backyard, and our yards adjoin. We each have pets too—I have a dog and you have a goose, and our pets like each other and often wander back and forth across the yards to play together.

One day, I discover that, while she was visiting with my dog, your goose laid an egg on my lawn. Upon inspection, I see that it isn't just any egg. It is a *golden* egg. My first thought, of course, is: *that's unusual.* But what if my second thought is: *that clever goose can hang out in my backyard anytime she'd like*—and I decide to keep the egg all for myself? After all, she did lay it in my backyard, and it is *solid gold*.

The next day, however, I find another golden egg in my grass. And the day after that I find another, and the day after that one more—and the day after that I start to realize this little goose is going to lay a golden egg on my lawn every day, regular as clockwork, and I have stumbled into a tidy little income stream. In this case, does it cross my mind that I ought to tell you about your goose's particular talent and cut you in on the golden egg deal? After all, it is your goose. You're the one paying for her goose food, and the upkeep on her cozy coop, and her vet bills, so shouldn't you benefit from the miraculous eggs she produces? If I do cut you in, though, what would you say would be a fair return on your investment in the goose? Just enough to cover your goose-related expenses, which you were going to have to cover anyway? Half, so you can make a profit on your pet just as I am doing? Certainly not all of it, because, after all, she is choosing to lay her eggs on *my* lawn….

You see where I'm going with this? To Big Tech, you are the goose, and your private data is all the golden eggs. Big Tech is making massive profits every hour of every day by selling your eggs, and mine, often when we don't even know they're doing it. They're creating potential problems and expenses for us all along the way—from the personal, like identity theft, to the internationally problematic, like crumbling democratic infrastructures. And they certainly aren't cutting any of us in on the profits.

But, if they did cut us in, what would you say would be a fair return for investing these behemoths with access to your private information?

Kara Swisher put it this way: "Since I first met them as struggling entrepreneurs, people like Elon Musk, Mark Zuckerberg and Jeff Bezos have become the wealthiest people in the history of the world by an enormous factor. What they have created is as impressive as it is immense. It is no less important than the invention of the printing press or electricity or the car. In fact, it is more so. But the consequences have also been dire, *and far too often we have been made into the fodder for all this so-called innovation.*"[371] (Emphasis mine.)

For the duration of this chapter, I ask you to put aside the real and urgent questions around the moral and legal rights of Big Tech to do whatever they like to do with our personal information, and I ask you to put aside even speculation about the solutions we really do have to the gritty problem of reining in Big Tech and policing how and when our information is gathered and used, and to what purpose. I'm asking you to focus on the nearly unimaginable wealth these tech titans have built by using us—using our private and intimate information—as fodder. As their own, personal supply of golden eggs. Focus here strictly on the

---

[371] Kara Swisher, "Dear Class of 2021: Don't Do Your Homework. Live Your Life." *New York Times* (June 28, 2021), https://www.nytimes.com/2021/06/28/opinion/covid-graduation-apple-steve-jobs.html.

financial stake each of us should have in the profits Big Tech is making off of us. It is our information that is creating their fortunes and wreaking havoc for us in the process, so shouldn't we have a piece of it?

○ ○ ○

The word *dividend* derives from the Latin word *dividendum*, which means a thing to be divided. A company's profits are a thing to be divided among the people who hold shares in that company, and so those shareholders are paid a dividend. A *data dividend* is an assignment of an economic value to our personal data, and an acknowledgment of the ownership of that value, as we would have in the case of any other property—say, a clever little goose.

I am not the first to propose data dividends as the solution for compensating those of us whose personal information has created such massive wealth for others. California, through its governor, Gavin Newsom, was at the forefront of calling for an accounting and payout for its citizens. In 2018, the state passed the landmark California Consumer Privacy Act (CCPA), a law that provides its citizens with "the right to know about the personal information a business collects about them and how it is used and shared; the right to delete personal information collected from them (with some exceptions); the right to opt-out of the sale of their personal information; and the right to non-discrimination for exercising their CCPA rights. Businesses are required to give consumers certain notices explaining their privacy practices. The CCPA applies to many businesses, including 'data brokers.'"[372] It is another state law—Civil Code section 1798.99.80—that defines the new career trajectory "data broker" as "a business that knowingly collects and sells to third parties the personal information of a consumer with whom the business does not have a direct relationship."[373]

But who are these companies, these data brokers? What exactly are they collecting, and to whom are they selling their collections?

---

[372] California Consumer Privacy Act (CCPA) website, https://www.oag.ca.gov/privacy/ccpa#sectiona.

[373] Ibid.

Julie Brill was nominated by President Barack Obama and confirmed unanimously by the Senate to serve a six-year term as Commissioner of the Federal Trade Commission. She now leads Microsoft's team that works on digital regulatory issues—and she confirms that you could reasonably call the information the data brokers collect about us a *dossier*. A dossier that not only contains our private information but that can be directly tied to each of us. "That's the whole point of these dossiers. It is information individually identified to an individual or linked to an individual."[374]

The data in those dossiers includes a trove of our most private facts and figures: our names, user names, religion, ethnicity, income, political affiliations, sexual orientation, and medical histories. Have you ever suffered from depression? It's in there. Do you take regular medications? The list is in your file. Have you ever sought treatment for drug- or alcohol-related problems? Duly noted in your dossier. One of the largest data brokers in the country is a company called LiveRamp, formerly Acxiom, which "brags it has, on average, 1,500 pieces of information on more than 200 million Americans."[375] The information that LiveRamp holds, *your* information, is then sold to advertisers, governments, or even other data brokers—and for a pretty penny: LiveRamp's total revenue for the second quarter of 2021 was $105 million, and they project a $113 million revenue for the third quarter.[376]

Not that LiveRamp is alone in the industry, of course. Tim Sparapani—a technology policy lawyer, privacy expert, and founder of SPQR Strategies—reports that "data brokers have been flying under the radar for years, preferring that people know as little as possible about

---

[374] Steve Kroft, "The Data Brokers: Selling Your Personal Information," *CBS News* (March 9, 2014), https://www.cbsnews.com/news/the-data-brokers-selling-your-personal-information/#textThe20largest20data20broker20is20Acxiom2C20a20marketingharder20for20Americans20to20get20information-20on20Acxiom.

[375] Ibid.

[376] LiveRamp Second Quarter Results, https://s22.q4cdn.com/928934522/files/doc_financials/2021/q2/Q2-FY21-Earnings-Press-Release_FINAL_110920_10AM.pdf.

the industry and the information that's being collected and sold. But the evidence is there if you know where to look."[377] Look in Danbury, Connecticut, for a company called Statlistics[378] if you're looking to market to gay and lesbian adults; to Paramount Direct Marketing[379] in Erie, Pennsylvania, for a list of people who suffer from alcohol, sexual, and gambling addictions; to Exact Data[380] out of Chicago, Illinois, if you want the names of people who've had sexually transmitted diseases, or who have purchased sex toys or other adult materials.[381] There is, perhaps alarmingly, almost nothing you can do online that won't get you on one list or another.

That includes how you travel around a city in the midst of a public health crisis. We've already talked about how law enforcement has used the GPS function installed in smart phones to track criminals. Well, in 2020, as governments all over the world were trying to figure out how to track the spread of COVID-19, the D.C. Department of Health entered into a "free trial"[382] agreement with the data brokerage Veraset to provide to them raw phone location data the broker "pitched as uniquely valuable for tracking the coronavirus pandemic".[383] While the District declined to continue using Veraset's services at the end of the free trial period—and while records obtained by the Electronic Frontier Foundation (EFF) through a Freedom of Information Act request contained no evidence that the data Veraset had collected had been misused—Bennett Cyphers, a technologist with EFF, suspects that the company "covid-wash[ed]"[384]

[377] Kroft, ibid.

[378] Statlistics website, https://www.statlistics.com/about-us.html.

[379] Paramount Direct Marketing website, https://www.paramountdirectmarketing.com.

[380] Exact Data website, https://www.exactdata.com.

[381] Kroft, ibid.

[382] Drew Harwell, "Data broker shared billions of location records with District during pandemic," Washington Post (November 10, 2021), https://www.washingtonpost.com/technology/2021/11/10/data-broker-shared-billions-phone-location-records-with-dc-government-part-covid-tracking-effort/.

[383] Ibid.

[384] Ibid.

the work they did during the pandemic. "'A lot of these data brokers' existence depends on people not knowing too much about them because they're universally unpopular,' Cyphers said. 'Veraset refuses to reveal even how they get their data or which apps they purchase it from, and I think that's because if anyone realized the app you're using' also 'opts you into having your location data sold on the open market, people would be angry and creeped out.'"[385]

There is legislation that addresses this problem. The California Consumer Privacy Act, for example, became effective on January 1, 2020, and enforcement of the new law began on July 1, 2020. But Caitlin Fennessy, research director at the International Association of Privacy Professionals (IAPP), points out that the CCPA does not focus on whether it's legal for a business to process an individual's information— as does the E.U.'s GDPR, which we will talk more about in upcoming pages—but only on "providing consumers with the ability to control whether their information is sold."[386] The penalties for violating the law are steep—$2,500 for each violation of the law and $7,500 for each intentional violation of the law. But it is not the companies that collect our information that must be proactive about staying within the law; rather, it is a matter of the consumer discovering that her or his rights have been violated and taking action against the offender.

While the struggle to pass the CCPA was still ongoing, the California legislature took up another key piece of related legislation—creating a data dividend that would pay Californians actual money for the use of their private data. Even back then, the idea of data dividends was not a new one. As reported by the *Los Angeles Times*, "The concept has been discussed in Silicon Valley for years as a way to tackle growing income

---

[385] Ibid.

[386] LJ Davids, " CCPA Deep Dive: How California is Enforcing its Major Privacy Law," *Security Management* (December 2020), https://www.asisonline.org/security-management-magazine/articles/2020/12/ccpa-deep-dive-how-california-is-enforcing-its-major-privacy-law/.

inequality at a time when tech companies have flourished and advances in automation have eliminated jobs."[387]

"California's consumers should also be able to share in the wealth that is created from their data," Newsom said at the time[388] about his ambitious plan. To date, though, no more concrete steps have been taken to make it a reality, mostly because figuring out how to calculate what a person's actual net data worth might be is a daunting task. We'll discuss the nuances for creating a formula that could enable those calculations later in this chapter. For now, let's continue to focus on the subsequent efforts of other states, and some individuals, to assert similar property rights over private information and encode those rights as laws.

For example, The Data Dividend Initiative is "an ad-hoc group of scholars and practitioners that formed to study and discuss the 'data dividend' concept."[389] The Data Dividend Project, an offshoot of Governor Newsom's work and led by former presidential candidate Andrew Yang, launched in 2020, aiming to do more than just study the issue. It is a movement that took shape around the goal of fulfilling CCPA's mandate. Based in Los Angeles, California, the organization is "dedicated to helping Americans regain ownership and control of our personal data. Our data is our property, and we should have the choice to get privacy and to get paid. We aim to mobilize millions of Americans to collectively exercise their data rights and to get paid if they choose to share their data."[390] The organization offers an online tool to help people whose privacy has been abused to ascertain whether or not they are eligible under the California law to file a claim, and then acts as a representative to help them to actually file, among other valuable services.

---

[387] Jazmine Ulloa, "Newsom wants companies collecting personal data to share the wealth with Californians," *Los Angeles Times* (May 5, 2019), https://www.latimes.com/politics/la-pol-ca-gavin-newsom-california-data-dividend-20190505-story.html.

[388] Ibid.

[389] The Data Dividends Initiative website, https://www.datadividends.org.

[390] Ibid.

In New York, State Senator David Carlucci hoped to follow California's lead, proposing in December of 2019 a 5 percent tax on Google, Facebook, and other tech companies that use New Yorkers' data. The funds would be collected by the state and delivered to citizens as an annual data dividend.[391] The bill, however, had trouble garnering sponsors and, to date, remains an idea only. The state that really did follow in California's footsteps, becoming only the second in the union to take concrete action to protect the online privacy of its citizens, was Virginia. There, in March 2021, Governor Ralph Northam signed into law the Virginia Consumer Data Protection Act (VCDPA). It becomes effective on January 1, 2023,[392] though it's important to point out that no one is getting money back yet, as there isn't a dividend provision per se within the act. It should also be noted that in April 2021, Virginia also banned the use of facial recognition technologies by police departments without first securing legislative approval.[393]

There have been other ongoing developments on the legal if not the legislative front, many of them positive, in terms of securing privacy for tech users. In Florida, news outlets may soon be able to sue social media platforms if their content is removed.[394] In Arkansas, Amazon must disclose contact information of third-party sellers to the state's

---

[391] Patrick Gleason, "Blue State Politicians Could Undermine Pushback Against EU Digital Taxes," *Forbes* (January 30, 2020), https://www.forbes.com/sites/patrickgleason/2020/01/30/blue-state-politicians-undermine-push-back-against-eu-digital-taxes/?sh=1f7079eb1014.

[392] "And Then There Were Two: Virginia Enacts Comprehensive Data Privacy Law," *The Data Dividend Project* (March 29, 2021), https://blog.datadividend-project.com/https-blog-datadividendproject-com-virginia-enacts-comprehen-sive-data-privacy-law-virginia-enacts-comprehensive-data-privacy-law/.

[393] Ibid.

[394] David McCabe and Cecilia Kang, "As Congress Dithers, States Step In to Set Rules for the Internet," *New York Times* (May 14, 2021), https://www.nytimes.com/2021/05/14/technology/state-privacy-internet-laws.html?referringSource=articleShare.

shoppers.[395] In February of 2021, the state of Maryland approved the country's first tax on Big Tech ad revenue.[396] And, speaking of facial recognition technology, in February 2021 a judge approved another settlement of a privacy lawsuit, this one against Facebook for allegedly using photo face-tagging and other biometric data without the permission of its users. The amount was $650 million, to be paid to almost 1.6 million Facebook users in Illinois who filed a class-action suit against the company in 2015. U.S. District Judge James Donato, who approved the deal, "called it one of the largest settlements ever for a privacy violation. 'It will put at least $345 into the hands of every class member interested in being compensated.'"[397] Additionally—and this part of the judgment is a win for all Facebook users interested in maintaining their online privacy—the company "agreed to set its face recognition default setting to 'OFF' globally."[398]

What you'll note though—in terms of actual cash exchanging hands and making it into the pockets of citizens who own the data that was involuntarily surrendered—is that this cash is the result of punitive rather than proactive measures. That is, the $650 million Facebook is paying to 1.6 million people in Illinois is the product of a nearly seven-year battle to hold the company accountable and punish it for misappropriation of private information—rather than an upfront recognition that the information holds intrinsic value in the marketplace; that the owner should have the option of voluntarily sharing it; and that said owner should be

[395] Arkansas State Legislature website, https://www.arkleg.state.ar.us/Bills/Detail?id=SB470&ddBienniumSession=2021%2F2021R.

[396] David McCabe, "Maryland Approves Country's First Tax on Big Tech's Ad Revenue," *New York Times* (February 12, 2021), https://www.nytimes.com/2021/02/12/technology/maryland-digital-ads-tax.html.

[397] "Judge Approves $650M Facebook Privacy Lawsuit Settlement," *ABC News* (February 26, 2021), https://abcnews.go.com/US/wireStory/judge-approves-650m-facebook-privacy-lawsuit-settlement-76150095.

[398] "Illinois Facebook Users to Receive $345 Each as Early as May 2021," *The Data Dividend Project* (March 4, 2021), https://blog.datadividendproject.com/illinois-facebook-users-to-receive-345/.

compensated fairly for doing so in a manner that's commensurate with its worth.

But what is it worth? What is *your* "'net worth"? What would be one individual's fair share of Big Tech's billions? A data dividend could be constructed as a percentage of a company's operating cash flow. That would generate a sizeable dividend to pay for everyone's data. In addition, if it is paid before taxes, these companies would have the ability to claim a tax deduction for the data dividend, which seems fair.

If we were to rely on this method of formulation, though, it would require a great deal of transparency from the company as well as legislating more uniform reporting methods—for example, breaking down tax reports to include such categories as ad revenue versus hardware sales, clearly defined and consistent across the board—so that a dividend payout would reflect income actually accrued through the use of private data. Even so, we'd still be left to determine a fair distribution based on any individual's contribution to the data pool. Should you, who works a job that requires you to spend eight hours a day online, receive the same dividend as your neighbor, who goes online for fifteen minutes a day only to quickly check her social media?

Or perhaps a more equitable distribution strategy would be to base dividends around a minimum wage—a sum paid for each hour a user spent in any various online activity, from social media interactions, to listening to music, to watching streaming television, to researching which doctor to choose in a new city, to placing an order for a grocery delivery. Would this result in more fairness in the dividend distribution model? Or would we, by placing a value on the time a user spent in front of a screen, only be inadvertently encouraging people to spend more time online?

Aside from the complex mathematics of the problem, several other factors add to the difficulty in creating a useable formula for an equitable data dividend. Data is very nearly an infinite asset. How does one go about putting a value on infinity? Lawmakers, for the most part, may be approaching solutions to the problem of data privacy and security with the best of intentions, but as a group they continue to demonstrate a basic lack of literacy around tech. It can feel, at times, as if we've tasked people who don't know how to fry an egg with making Eggs Benedict. For twelve. Over a campfire. And there is, in the United States, the unique situation that its laws and regulations are so very fragmented, state by state, that creating a unified resolution is, if not impossible, then frustrating, unwieldy, and far too slow.

These are among the reasons why the United States is lagging behind the E.U. in the arena of tech regulation. But if we cut to the core cause of the delay in action, we arrive at a particularly political impasse. Both sides of the American aisle, Democrats and Republicans, liberals as well as conservatives, are mad at Big Tech, but they are mad for diametrically opposing reasons. Democrats are angry that Big Tech has allowed misinformation and disinformation to flourish, unchecked, on their various platforms, exposing small 'd' democracy as well as the lives of America's citizens to risk and danger. Think of all the people still clinging to the falsehoods that Trump won the 2020 election, or that the COVID-19 vaccines are really just an excuse to plant a microchip in your arm, and you'll see their point. Republicans, conversely, think that in trying to stem the flow of misinformation and disinformation, Democrats are making an attempt to censor their right to free speech. Indeed, whole new social networks have either been proposed or actually sprung to life in response to this fear on the right. Take the case of Gettr, which emerged in the summer of 2021 with, as Politico reports, a "free speech policy" that will "purportedly... allow users to fully express themselves without the censorship of tech giants."[399] The problem with "free speech" as practiced by Gettr? Jihadists have flocked to the site to speak freely too, and so, Gettr "features reams of jihadi-related material, including graphic videos of beheadings, viral memes that promote violence against the West and even memes of a militant executing Trump in an orange jumpsuit similar to those used in Guantanamo Bay."[400]

---

[399] Mark Scott and Tina Nguyen, "Jihadists Flood Pro-Trump Social Network with Propaganda," *Politico* (August 2, 2021), https://www.politico.com/news/2021/08/02/trump-gettr-social-media-isis-502078.

[400] Ibid.

In a governmental atmosphere in which it is difficult to get both sides to come together and decisively conclude even that balmy spring weather beats a blizzard, you can zoom in like a Hubble telescope on a primary source of the gridlock around this issue.

Further, we have to consider that Big Tech isn't shy about spending money to lobby members of congress. Now, the idea of lobbying a legislator has taken on a fairly suspect meaning, but it isn't a nefarious practice in and of itself. Indeed, lobbying—defined as individuals or groups of individuals with a collective and particular interest in pressuring legislators to enact certain policy actions—is actually necessary in a participatory democracy like the United States. Every time you pick up the phone to tell your town councilman that the pothole in front of your house remains unrepaired, or write a letter to your mayor to encourage her to spend money to hire another policeman or two to patrol your city streets, or make a trip to your state capital or even Washington, D.C., to join a protest organized by a group in which you have an interest, you are, in fact, lobbying.

The problem with lobbying arises when large corporations, like the ones that make up the U.S. tech sector, put vast wealth of the sort that an *extremely* small percentage of real, living, breathing citizens have at their disposal at the service of their desired policy outcomes. For example, in the first half of 2020, when legislation was being debated around a coronavirus relief package, Facebook, Amazon, Apple, and Google spent a combined $20 million to lobby Congress on the issue.[401] How much Big Tech spends on lobbying efforts will—and *should*—astonish you. Big Oil and Big Tobacco were, once upon a time, the biggest spenders on lobbying efforts; but these days Amazon and Facebook alone spend

---

[401] Brian Schwartz, "Big Tech Spends Over $20 Million on Lobbying in First Half of 2020, Including on Coronavirus Legislation," *CNBC* (July 31, 2020), https://www.cnbc.com/2020/07/31/big-tech-spends-20-million-on-lobbying-including-on-coronavirus-bills.html.

nearly twice as much on lobbying as Exxon and Philip Morris.[402] How much is twice as much? In the 2020 election cycle, Big Tech spent $124 million in campaign contributions alone, breaking its own record from past cycles.[403] In the short span between 2018 and 2020, Amazon increased its spending on lobbying by 30 percent, and Facebook increased its spending by an incredible 56 percent.[404]

Put your well-intentioned telephone call to your congressional representative up against $124 million—heck, put the calls from you and ten thousand of your closest compatriots up against that figure—and you'll have a good glimpse into why the flurry of activity around tech in D.C. feels so futile, why "the nonstop barrage of high-profile CEO hearings and scathing reports has so far failed to produce any significant new law reining in tech's power."[405]

o o o

The European Union, on the other hand, has long been at the forefront of tech regulation and data privacy protection, and it announced in February 2020 the creation of a truly unified data strategy for its citizens.[406] Under "The Trusts Project," set to be implemented in 2022, a pan-European pool of information—both personal and non-personal—will be created for its nearly 500 million citizens. The information will be held on public servers, managed through the Trusts Project, and, importantly, global tech companies will no longer be allowed to accumulate this

[402] Jane Chung, "Big Tech, Big Cash: Washington's New Power Players," *Public Citizen* (March 24, 2021), https://www.citizen.org/article/big-tech-lobbying-update/.

[403] Ibid.

[404] Ibid.

[405] Scott Rosenberg, "The Techlash Is a Bust," AXIOS (May 2, 2021), https://www.axios.com/techlash-bust-big-tech-regulation-4f87efab-d35c-4d2d-abaa-dafcfe729118.html.

[406] "Communication from the Commission to the European Parliament, the Council, the European Economic and Social Committee and the Committee of the Regions," *European Commission*, https://ec.europa.eu/info/sites/default/files/communication-european-strategy-data-19feb2020_en.pdf.

information privately but will need to access it through the trusts—essentially making the trusts the clearinghouse for all personal information about E.U. citizens.

The Trusts Project is, of course, not without its critics. The "assetization" of personal data may end up pitting the interests of private enterprise against those of the public sector, creating conflicts of interest for those managing the data.[407] In India, the trusts structure defined its managers as "information fiduciaries," and thus gave the government unrestricted access to the personal information of its citizens.[408] While the intent of the E.U.'s project is for the E.U. itself to profit from the information of its citizens, it does intend to distribute dividends to its citizens in exchange for the data's use. But again, the formula and process for calculating an amount for those distributions is not clearly defined.

Some private companies are stepping into the void. For example, Folia Health[409] recognized the importance of collecting certain personal data for use in research purposes, and in order to encourage patients to opt in and allow the company to collect and analyze private data, it instituted a data dividend. Currently, their data dividends are valued at four dollars per month, paid by way of a Walmart, Amazon, or Starbucks gift card. Think about this, however: if the company is willing to voluntarily pay a patient four dollars per month to use his or her data, imagine how much that data is really worth to the company. While it's a small step, it is still a notable and proactive one.

In the most desirable and effective of scenarios, though, we would be able to create an equation through which each tech user would receive a share of the billions of dollars tech giants are amassing by using our data. I'll restate a point here, because it's an important one: creating an equitable formula is nearly impossible without transparency from Big Tech. Failing that, we have no way to either access or audit the numbers and

---

[407] "European Commission," ibid.

[408] Smith Krishna Prasad, "Information Fiduciaries," *Data Catalyst* (September 2019), https://datacatalyst.org/wp-content/uploads/2020/06/Information-Fiduciaries-and-Indias-Data-Protection-Law.pdf.

[409] "Data Dividends," *Folia Health*, https://www.foliahealth.com/datadividends.

statistics we would need in order to make the equation work. I'll move along optimistically in spite of that huge obstacle and outline the factors that ideally should go into the creation of such an equation:

- The number of searches a user conducts per day.
- The recency of the user's searches.
- The dollar value of what a user buys online in a given year.
- The user's income demographic.
- The amount of time a user spends online per day.
- The amount of time a user spends on a specific website.
- Whether or not the user has a smartphone.
- Whether or not the user has social media—this is because social media profiles allow brands to pinpoint potential customers more easily and, without social media, the user becomes more difficult to track.
- Where the user is located—meaning the cost to deliver an item to the user is dependent on whether the user lives near to the location of a warehouse or somewhere more remote.

Given these parameters, a formula for calculating the amount of an individual's data dividend check might look like this: $S+T+D+P+SM$ ($+$ or $-SP$) $= DD$. The key to the equation is:

S = the number of searches the individual conducts per day;

T = time the individual spends online per day, and on specific sites;

D = the individual's demographic, which would include her or his age and location;

P = purchase power, meaning how much the individual spends online per month;

SM = social media;

SP = whether or not the individual owns a smartphone, and is indicated with a simple plus or minus;

DD = your individual Data Dividend.

As you can see, this is no run-of-the-mill Average Revenue Per Unit (ARPU) calculation, which can be more easily arrived at by straightforwardly dividing revenue by the number of units, subscribers, or users. If an ice cream store netted $500 on Monday, based on the sale of 150 cones, that would mean the ARPU was $3.33 per cone. This is an equation that is elegant in its simplicity, and it isn't going to work at all in the context of calculating the value of any one person's data.

Indeed, getting the equation I've created to work for our purposes isn't going to be possible without the aforementioned transparency by Big Tech. For example, there has been no credible research conducted on the average number of searches any given individual might conduct over the course of a day—some sources say three to four searches, some say seven to ten. Anecdotally, as anyone who has ever lived with a young adult will attest—watching them nearly simultaneously Google their research for a homework assignment, peruse online menus for an after-school snack, and pull up the site of the local DMV office to find out when they can schedule their driver's test—they simply have to be raising the average for the rest of us.

Still, there are elements of simplicity in the data dividend proposal. The equation we're looking for is a way to analyze Big Tech's profit based on user data, and then proportionally tax Big Tech, ultimately using those tax dollars to return to the customer a fee for the use of the data. We're looking for a way to put some of the money Big Tech makes on the back of your personal information back in your pocket. Just as once a year most people get a check from the IRS, most people should also get one from Big Tech, reflecting payment for the use of her or his data.

But let me be clear: what we want to avoid is for the calculations of data dividends to become as complicated as IRS calculations. So, let's think about how to simplify even further. Let's focus on how much a data dividend might be worth to an individual user.

One well-placed, senior industry executive told me that, depending on how and how long the individual used his or her devices on average, their data dividend could be worth between seventy-five and one hundred dollars per month. It's an amount that won't make anyone rich, but we're not in the realm of pennies either. Perhaps the simplest, most elegant solution, then, would be to simply decide that each tech user should receive a flat rate—a dividend of, say, fifty dollars per month. As Big Tech has our credit card numbers already, the dividend could be credited directly to our accounts on the first of each month. In the case of users who don't have a credit card, or don't use them for online purchasing, these users would be issued an old-fashioned check.

My proposal also includes some parameters regarding distribution—who should be the recipients of those Big Tech tax dollars. While acknowledging the fact that higher-income people own data that is more valuable to Big Tech, it's also important to realize that, in our connected and well-tracked world, privacy itself is a new luxury good. Writing for the *New York Times*, correspondent Jill Cowan acknowledged the concern about issuing "a straightforward payment or reward for consumers in exchange for the use of their data. Though that's an idea that has raised concerns among privacy advocates who worry such a model could make privacy even more of a luxury, accessible only to those who can afford to opt out of selling their data."[410] Therefore, I propose we take our cues from guidelines established for COVID-19 Economic Impact Payments. That is, if your income is more than $198,000 and/or you don't "qualify as an adult,"[411] then you won't receive a data dividend check. I also propose that a portion of the tax received in this way from Big Tech be designated for digital literacy programs, and/or for making broadband more accessible to the nation's rural neighborhoods—and, importantly, school districts where it is now either absent or in short supply.

---

[410] Jill Cowan, "How Much Is Your Data Worth?", *New York Times* (March 25, 2019), https://www.nytimes.com/2019/03/25/us/newsom-hertzberg-data-dividend.html.

[411] In later rounds of stimulus check distribution, people as young as seventeen qualified as a adults—that is, were eligible to receive the checks—if they had filed a tax return in the previous year and met the government's definition of an eligible dependent.

o o o

All these approaches to creating a data dividend take for granted, however, the shared assumption that our data will in fact be gathered, stored, and used by Big Tech. In its Trusts Project proposal, the E.U. even positions the sharing of information as a civic duty,[412] but that isn't necessarily the only alternative. What if we—or you, or I, as individuals—don't want to share our private information with Big Tech? What happens in that case? Let's now turn to look at what we can do to protect our data from Big Tech's prying eyes in the first place.

---

[412] "European Commission," ibid.

Chapter 6

Power to the People, Solutions for a Digital Democracy

───○─○─

The January 6, 2021 attack on the Capitol was not a surprise. "On January 5th, the FBI shared intelligence with the U.S. Capitol Police of the upcoming threat with the following instructions, 'to not take action based on this raw reporting without prior coordination with the FBI.'"[413] In reality, many of us could see the insurrection coming, even if we might not have verbalized it as such before the shock of seeing an actual insurrection happening in real time on our television screens. Yale historian, Timothy Snyder, in the context of discussing the impact of Russia's actions on the 2016 election at the American Academy in Berlin in 2019, sketched out three main objectives of intelligence work in this way: "intelligence" is the process of gathering of information; "counterintelligence" is the process of keeping the other guy from finding out information your side wants to keep to itself; "active measures" "are when you look for the psychological weakness... of your opponent and

───────────

[413] Devlin Barrett and Matt Zapotosky, "FBI report warned of 'war' at Capitol, contradicting claims there was no indication of looming violence," *Washington Post* (January 12, 2021), https://www.washingtonpost.com/national-security/capitol-riot-fbi-intelligence/2021/01/12/30d12748-546b-11eb-a817-e5e7f8a406d6_story.html.

you try to get your opponent to do the things that aren't in his interest for reasons that he can't quite understand."[414] The public could see the provocation to action in slogans ("save America" and "stop the steal"), in incendiary speeches at the rally that immediately preceded the attack on January 6th, and, as the *New York Times* put it, Donald Trump "all but circled the date" with his December 19, 2020, tweet about the 6th: "Be there, will be wild!"[415]

Indeed, in many other ways too, the attack on the Capitol was a fairly transparent effort: it was organized on social media. According to the Tech Transparency Project (TTP), the attack was months in the making, particularly on Facebook—no matter that Facebook Chief Operating Officer, Sheryl Sandberg, insisted that the mob action against the U.S. Capitol and sitting U.S. elected officials was "largely organized" on other platforms.[416] "Not only is that assertion false, according to research by the Tech Transparency Project (TTP), but it ignores the fact that Facebook spent the past year allowing election conspiracies and far-right militia activity to proliferate on its platform, laying the groundwork for the broader radicalization that fueled the Capitol insurrection in the first place."[417]

Much of the organizing took place on Facebook *private groups*, in which only other members can see who is in the group and what those other members are posting. Creating a private group on Facebook is quite a simple task: just click on the dropdown "create" menu located, as of this writing, in the top right quadrant of your Facebook page, and indicate that you'd like to create a group. Fill out the information that Facebook requests—the name of the group; a brief statement concerning

---

[414] "The Road to Unfreedom: Democracy, Neofascism, and the Importance of Language," *YouTube* (April 3, 2019), https://www.youtube.com/watch?v=VHDdzJXM4oY.

[415] Dan Barry and Sheera Frenkel, "'Be There. Will Be Wild!': Trump All but Circled the Date," *New York Times* (January 6, 2021), https://www.nytimes.com/2021/01/06/us/politics/capitol-mob-trump-supporters.html.

[416] "Capitol Attack Was Months in the Making on Facebook," *Tech Transparency Project*, https://www.techtransparencyproject.org/articles/capitol-attack-was-months-making-facebook.

[417] Ibid.

what the group is about; the privacy setting of the group, meaning, will it be restricted only to those you invite to participate, or can anyone who is interested in its subject request membership?—then invite your friends and other interested parties to join, and *voilà*, you have a private group. Alternatively, in lieu of starting your own, you can request to join a private group that already exists, and there are hundreds of thousands of such groups, if not millions. There are private groups based around families, so members can stay abreast of the celebrations and milestones in the lives of other family members; there are alumni groups, so those who graduated together can stay in touch; there are support groups for those caring for elderly parents or a child who isn't neurotypical; there are information groups centered around those who have seed allergies; there are groups that are essentially fan clubs for pop stars and sports figures and even politicians. And there are politically oriented groups, like the 23,000-member one called "The Patriot Party" that shared this since-deleted post in December of 2020: "If they won't hear us, they will fear us. The Great Betrayal is over."[418]

Don't get the wrong idea—it wasn't *only* Facebook that allowed such incendiary posts. Twitter, Reddit, YouTube, Gab, and Parler also hosted their fair share of the sedition-minded. "Over the last year, millions of right-wing social media users have quickly been indoctrinated into online extremist movements—largely thanks to vague terms and conditions put in place from the social media companies that play host to their communication."[419]

Renee DiResta, a researcher at the Stanford Internet Observatory who studies online movements, said, about the January 6th violence that took place on behalf of Donald Trump: "These people are acting because they are convinced an election was stolen. This is a demonstration of the

---

[418] David Mack, Ryan Mac, Den Basinger, "'If They Won't Hear Us, They Will Fear Us': How the Capitol Assault Was Planned On Facebook," *BuzzFeed News* (January 19, 2021), https://www.buzzfeednews.com/article/davidmack/how-us-capitol-insurrection-organized-facebook.

[419] Julia Sachs, "The Domino Effect: How Social Media Staged a Coup," *Grit Daily* (January 7, 2021), https://gritdaily.com/coup-twitter-jan-6/.

very real-world impact of echo chambers."[420] She added: "This has been a striking repudiation of the idea that there is an online and an offline world and that what is said online is in some way kept online."[421]

Joan Donovan, Research Director of the Harvard Kennedy School's Shorenstein Center on Media, Politics, and Public Policy, frames the convergence of offline and online life in this way: "As Donovan sees it, governments need to regulate and oversee the digital space because online activities have real-world effects. 'I think everybody knows that people form communities online, and it's in those bonds and in those spaces that people make decisions....'"[422] That is, people do not make a natural distinction between the group they belong to on Facebook and their real life: everybody else in the "patriot" group is planning to spend the time and money to go to Washington, D.C. to "stop the steal," and they are *my people*, so why shouldn't I plan on going to D.C., too?

One might have thought that the insurrection would force lawmakers to get serious about regulating Big Tech and move speedily toward solutions—that the provocation to the violence that occurred at the Capitol would be among the myriad issues to be addressed, information to be brought to light, and problems to be solved in the wake of the mob attack. However, it was social media platforms themselves, perhaps surprisingly, that reacted with more alacrity than lawmakers. Among the actions Big Tech took was that Twitter, Facebook, and even Pinterest suspended the then-president's accounts, effectively silencing a main source of the incendiary disinformation that had led to the violence.

And the lawmakers? Although the social media aspect of the violence was only one among the mass of security issues associated with

---

[420] Sheera Frenkel, "The storming of Capitol Hill was organized on social media.", *New York Times* (January 6, 2021), https://www.nytimes.com/2021/01/06/us/politics/protesters-storm-capitol-hill-building.html.

[421] Ibid.

[422] Joan Donovan, "On How Platforms Enabled the Capitol Hill Riot," Centre for International Governance Innovation (January 21, 2021), https://www.cigionline.org/big-tech/joan-donovan-how-platforms-enabled-capitol-hill-riot/.

the insurrection,[423] lawmakers, as a body, did not respond to the event with the same urgency as had the social media platforms themselves. In March of that year, the House Energy and Commerce Committee roundly scolded the CEOs of Facebook, Twitter, and Google: "'You're not passive bystanders,' said Rep. Frank Pallone (D-N.J.), chairman of the committee. 'When you spread misinformation, actively promoting and amplifying it, you do it because you make more money.'"[424]

One month later, in April, the Senate Judiciary Subcommittee on Privacy, Technology and the Law held bipartisan hearings at which they grilled representatives from Facebook, YouTube, and Twitter, as well as "researchers who testified that the algorithms used by the platforms to generate revenue by keeping users engaged pose existential threats to individual thought, and democracy itself."[425] One of those was Tristan Harris, a data ethicist who runs the Center for Humane Technology. His response to the representatives from Facebook, YouTube, and Twitter testifying before the committee about their operations? "It's almost like having the heads of Exxon, BP, and Shell here and asking about what you're doing to responsibly stop climate change. Their business model is to create a society that's addicted, outraged, polarized, performative and disinformed."[426] The next month, in May 2021, nearly five months after the riot, a bill to create an independent inquiry to investigate the insurrection—an inquiry that would likely have revealed the way that social media was used, and by whom, on January 6th—was voted down 54–35

---

[423] James Z. Boykin and Frederick Paige, "What Security Lessons Did We Learn from the Capitol Insurrection?" *Brookings* (March 10, 2021), https://www.brookings.edu/blog/fixgov/2021/03/10/what-security-lessons-did-we-learn-from-the-capitol-insurrection.

[424] Ina Fried, "Congress tells tech CEOs to buckle up for new laws," AXIOS (March 26, 2021), https://www.axios.com/newsletters/axios-login-950953c7-ec6d-4280-86a7-8b5206cde00f.html.

[425] Dean DeCharo, "Social Media Algorithms Threaten Democracy, Experts Tell Senators," *Roll Call* (April 27, 2021), https://www.rollcall.com/2021/04/27/social-media-algorithms-threaten-democracy-experts-tell-senators/.

[426] Ibid.

in the U.S. Senate.[427] It wasn't until July 1, 2021—nearly seven months after the insurrection—that the U. S. House of Representatives voted to create the United States House Select Committee to Investigate the January 6th Attack on the United States Capitol, and not until September 23, 2021, that the Committee issued its first round of subpoenas.[428]

What did move lawmakers to take a nearly immediate measure of action was the hack of the Colonial Pipeline that spans New Jersey to Texas and is responsible for delivering nearly 50 percent of the gasoline used on the East Coast of the United States. This ransomware attack occurred on May 7, 2021. "In a ransomware attack, hackers seize control of a business or organization's computer system by exploiting weaknesses in the security system, then lock up the entire system until a 'ransom' is paid."[429] In this case, the hackers did not manage to take control of the operations of the pipeline, but they did lock the company's computer systems, demand a ransom of $5 million, and cause gas shortages in, notably, North Carolina, Georgia, and Washington, D.C., where up to 65 percent of area gas stations experienced outages.[430]

In response, Colonial Pipeline paid the ransom,[431] and on May 12, 2021, President Joe Biden issued an Executive Order, a framework for the

---

[427] Ryan Nobles, Ted Barrett, Manu Raju and Alex Rogers, "Senate Republicans Block January 6 Commission," *CNN* (May 28, 2021), https://www.cnn.com/2021/05/28/politics/january-6-commission-vote-senate/index.html.

[428] Jacob Crosse, "Select Committee Issues First Round of Subpoenas Targeting High-Level Trump Conspirators in January 6 Coup Attempt," *World Socialist Web Site* (September 24, 2021), https://www.wsws.org/en/articles/2021/09/25/coup-s25.html.

[429] David Cohen, "Ransomware attacks 'are here to stay,' Commerce secretary says," Politico (June 6, 2021), https://www.politico.com/news/2021/06/06/ransomware-attacks-commerce-secretary-492005.

[430] Will Englund and Ellen Nakashima, "Panic buying strikes Southeastern United States as shuttered pipeline resumes operations," Washington Post (May 12, 2021), https://www.washingtonpost.com/business/2021/05/12/gas-shortage-colonial-pipeline-live-updates/.

[431] Eamon Javers and Amanda Macias, "Colonial Pipeline Paid $5 Million Ransom to Hackers," *CNBC* (May 13, 2021), https://www.cnbc.com/2021/05/13/colonial-pipeline-paid-ransom-to-hackers-source-says.html.

need to, among other measures, remove barriers among agencies sharing threat information, modernize the U.S.'s cyber security, and establish a Cyber Safety Review Board.[432] Hackers responded to these actions with a cyberattack on JBS SA, the world's largest meat producer, shutting down all of its U.S. beef plants, "wiping out output from facilities that supply almost a quarter of American supplies."[433] Although, in June, the FBI recovered $2.3 million of the Colonial Pipeline ransom paid in bitcoin to the Russia-based hacking group called DarkSide,[434] Commerce Secretary Gina Raimondo warned the business community that ransomware attacks "are here to stay."[435] She also took the opportunity to add that one way to fight back against ransomware attacks would be to support the Biden administration's American Jobs Plan, because "certain components of the [plan] provide for investments to shore up the nation's cyber infrastructure"[436]—though, given the sharp partisan divides that exist in Washington, D.C., how likely it is that the plan makes it into law, or makes it intact, is a big question mark.

o o o

It is clear that we have not yet figured out how to live together amicably in our new "social media city," as it was framed by Sahar Massachi and Kathy Pham in a podcast for the Berkman Klein Center for Internet &

---

[432] "Executive Order on Improving the Nation's Cybersecurity," *The White House* (May 12, 2012), https://www.whitehouse.gov/briefing-room/presidential-actions/2021/05/12/executive-order-on-improving-the-nations-cybersecurity/.

[433] Fabiana Batista, Michael Hirtzer, and Mike Dorning, "All of JBS's U.S. Beef Plants Were Forced Shut by Cyberattack," *Bloomberg* (May 31, 2021), https://www.bloomberg.com/news/articles/2021-05-31/meat-is-latest-cyber-victim-as-hackers-hit-top-supplier-jbs.

[434] Nicole Sganga, "U.S. Recovers $2.3 Million In ransom Paid to Colonial Pipeline Hackers," *MSN* (June 8, 2021), https://www.msn.com/en-us/news/us/us-recovers-2423-million-in-ransom-paid-to-colonial-pipeline-hackers/ar-AAKNRf7.

[435] Cohen, ibid.

[436] Ibid.

Society at Harvard University.[437] Massachi and Pham posit that, in addressing a range of problems associated with social media, such as how these platforms can amplify harassment, hate speech, and violent extremism, we adopt a new problem-solving model and think of social media as a city, in order to create a successful and humane way to govern it.

This is a hopeful way to think about our online lives. It suggests the shared responsibility of community—the best of what technology might bring us if we can appropriately manage it. Less hopeful is that the people we have put in charge of managing it for us are lawmakers who are, in large part, in over their heads in terms of knowing how the various technologies and platforms actually work—or the techno folks themselves who are making massive amounts of money in the as-yet-unregulated cyber frontier and will only continue to profit by maintaining the status quo. For example, most of Congress is made up of lawyers, business people, or career politicians, with only about 10 percent of members having a background in science, technology, engineering, and mathematics (STEM)—and it is people with STEM backgrounds who could be most helpful in addressing contemporary and emerging issues such as cybersecurity, information technology, and privacy concerns.[438] In fact, the need for expertise on Capitol Hill as legislators grapple with tech issues is so strong and dire that an organization, The Day One Talent Hub, was formed for the express purpose of helping "federal agencies to identify and access world-class technical talent to advance progress on priorities where science and technology can make a dramatic impact."[439]

Even less hopeful is the simple fact that tech is always ahead of regulation by a good seven to ten years, so the lawmakers have a great deal of catching up to do even as tech keeps racing forward.

---

[437] Sahar Massachi and Kathy Pham, "Governing the Social Media City," *Berkman Klein Center* (April 20, 2021), https://cyber.harvard.edu/events/governing-social-media-city.

[438] "Membership of the 117th Congress: A Profile," *Congressional Research Service* (updated August 5, 2021), https://fas.org/sgp/crs/misc/R46705.pdf.

[439] Day One Project website, https://www.dayoneproject.org/talenthub.

The first step toward solving any problem is, of course, awareness of it. We can't imagine how to fix a thing if we don't recognize that it's broken in the first place. In this chapter, we're going to imagine what actions we can take—individually as well as collectively—toward solutions that will make for a safe, fair digital landscape. What would it take to reform the ways we relate to and use technology? What do we have to do *now* to structure an online life that will provide the most benefits in the future, and minimize the downsides? What would a social movement based around proactively shaping—and *re*shaping—our contemporary digital landscape look like? This is a big subject that covers a great deal of territory, so settle in for a provocative journey along the cyber highway.

○ ○ ○

Let's begin with a foundational concept: *net neutrality*. This is a phrase that has been kicking around for years, not always precisely defined and—perhaps because its parameters were so permeable—yawn-inducing. In this chapter, however, we're going to approach each facet of each digital issue by asking direct questions, so we can seal up the leaks in our understanding of tech and find simplicity, and even elegance, in solutions. By definition, net neutrality is "the idea, principle, or requirement that internet service providers should or must treat all internet data as the same regardless of its kind, source, or destination."[440] But what does this mean in practice?

Right now, we are used to the internet as a free and open place, meaning that, in theory, every domain is given equal weight in terms of accessibility. For example, when you're looking to buy a print copy of the latest thriller from John Grisham, you can get online and shop as easily at your local bookstore as you can at Amazon. Both websites load at equal speed and, depending on the ecommerce configuration of your local store's site, you can check out just as seamlessly on either site. Perhaps it's important to you to support your local businesses, or maybe you want to

---

[440] Merriam-Webster website, https://www.merriam-webster.com/dictionary/net%20neutrality.

read the book tonight and you don't want to wait even a day for delivery, so you'd rather drop by the local place to pick up your copy. Whatever reason you may have for preferring to patronize a business in your town doesn't really matter because, right now, you have an accessible choice: the internet is *neutral*; it doesn't play favorites by making one site harder to use than another.

In December of 2017, then-President Trump's FCC voted to do away with such neutrality, on the grounds that broadband network investment "had tanked"[441] since net neutrality rules had taken effect two years before—and in spite of industry reports that they had "found…not a single publicly traded US ISP ever told its investors (or the SEC) that [net neutrality] negatively impacted its own investments specifically."[442] What this repeal of net neutrality means is that those businesses that can afford to pay a higher fee for premium access—faster speeds and better connections, as examples—will offer better, faster service to consumers. In plain words, the smaller businesses, like your local bookstore, will be squeezed out, profits for the behemoths will continue to grow, and their monopolies will become even more consolidated. Make no mistake about it, Big Tech very much likes the idea of abandoning net neutrality. In a report released in May 2021, "New York Attorney General Letitia James found that of the more than 22 million public comments the Federal Communications Commission received in 2017 regarding the repeal of net neutrality protections, a whopping 18 million of them—an astounding 82%—were fake. Millions of those comments, the report says, were funded by the broadband industry."[443]

---

[441] Jon Brodkin, "Title II Hasn't Hurt Network Investment, According to the ISPs Themselves," *ARS Technica* (May 16, 2017), https://arstechnica.com/information-technology/2017/05/title-ii-hasnt-hurt-network-investment-according-to-the-isps-themselves/.

[442] Ibid.

[443] Issie Lapowsky, "NY AG Finds Nearly 82% of Net Neutrality Comments to the FCC Were Fake," *Protocol* (May 6, 2021), https://www.protocol.com/fcc-net-neutrality-fake-comments.

What can we expect if net neutrality goes away? Small, local stores with a smaller online presence—and smaller budgets than the big guys to boost that online presence—will be unable to compete. Their doors will close. The landscapes of small as well as large city downtowns have been changed, likely forever—changes brought on by the dual forces of internet shopping and pandemic closures. If small businesses are going to be able to sustain themselves as we make the transition to an increasingly digital economy, maintaining net neutrality is not negotiable.

The good news is that, in spite of Trump administration rulings, the fight to keep the net neutral isn't over. There are any number of lawmakers who are committed to a neutral net, and dedicated lobbying efforts to hold Congress's feet to the fire. Though net neutrality isn't only an American issue, of course; it's a global concern. It's critical that we look for solutions to the problems associated with Big Tech through a global lens—and for two overarching reasons.

First, as we move ever more deeply into a digital economy—that is, as more and more commerce takes place through digital mediums, from shopping to banking, food service to travel arrangements—borders become ever more blurred. I can order cashew chicken salad from my favorite local restaurant or my favorite Japanese noodles direct from the shop where they're made in Tokyo—and have both delivered direct to my door. You can deposit your paycheck in your local bank branch in Michigan even while you're on vacation in Maine—and you can wire extra funds from the desk in your home office in the United States to your kid who's traveling in Mexico without even getting out of your chair. The convenience is astonishing, though such seamless ease—such "frictionless" commerce—has long been the goal of Big Tech and is now so much a part of daily life that, most of the time, we forget to be astonished about it.

With convenience and ease for us, however, come complications for business and for the governments under which rules they conduct commerce. Notably, these include complicated accounting procedures. For example, you may have a U.S.-based company that sells its goods or services online, but *where* you sell those goods and services can create a mire come tax season. In what countries do your customers live? What are that country's tax laws, and its tax rate? When are those taxes due? This is a dramatically simplified list of questions corporate accountants would have to ask on behalf of their employers in a significantly less complex and more transparent global tax environment. These are questions that become ever more relevant as we move more deeply into the digital age, and a company no longer has to have a large, brick-and-mortar presence in a country in order to make profits within that country.

On June 5, 2021, world leaders took a meaningful and unprecedented first step toward creating a less chaotic tax environment, one that is more relevant in our global digital economy. The Group of Seven (G-7)—comprised of Canada, France, Italy, Germany, the U.K., Japan, and the United States—is an intergovernmental organization that meets from time to time to address key international economic and monetary issues. At the G-7's June 2021 meeting in London, treasury chiefs agreed to back a global minimum tax rate of 15 percent for companies that have a profit margin of at least 10 percent.[444] Their backing doesn't mean that this *is* the rule as of June 2021. The measure must be approved next by the G-20—a group of finance ministers and bank governors from nineteen countries and the E.U., representing the world's largest economies and charged with the regulation of financial markets and promoting global economic growth. Trade agreements will have to be renegotiated to reflect the new parameter. This will not be an easy process or a quick one, but "the G-7 agreement brings a possible increase in tax bills for a number of digital businesses a step closer."[445]

Why is this an important first step for consumers? Well, it means an increase in national tax revenues, for a start. As of this writing, all five of the Big Tech companies we are primarily focused on are incorporated in the United States: Google and Facebook in Menlo Park, California;

---

[444] Paul Hannon, Richard Rubin and Sam Schechner, "G-7 Nations Agree on New Rules for Taxing Global Companies," *The Wall Street Journal* (June 5, 2021), https://www.wsj.com/articles/g-7-nations-agree-on-new-rules-for-taxing-global-companies-11622893415?st=z97jfuo5ktme6i7&reflink=article_email_share.

[445] Ibid.

Apple in Cupertino, California; and Microsoft and Amazon in the state of Washington. Also, the move would do away with the incentive for a company to uproot itself from the country in which it is currently based and relocate to one that has a more favorable tax rate. That is, it would do away with tax havens and provide for corporations to pay their fair share of taxes within the country where they are headquartered, and whose public services and infrastructure—such as transportation networks and police forces—that company and its employees profit from using. At the same time, negotiations for a more uniform minimum global tax rate provide an entry point at which to open the discussion of creating a digital dividend for consumers, whose private information forms the foundation upon which those global corporate profits are built—perhaps on the order of what the E.U. has already proposed in its Trusts Project. At least one industry insider estimates that such a dividend could be worth up to seventy-five to one hundred dollars per individual user, per month.

The prospect of using ideas already outlined in the E.U. Trusts Project brings us neatly to the second reason it is important to maintain a global perspective while creating solutions to digital problems. The reason is that we—meaning the United States—can learn a great deal from countries and organizations that have been at work on the problem a lot longer and more intently than we have been. Why are other countries so far ahead of the United States when it comes to imagining the future of tech and regulating it? The answer is layered, though a large part of it can be summed up in an understanding of the unique relationship between the country and the industry.

First of all, Big Tech doesn't control European countries in the same way they control the U.S. government, and this is because tech companies are, in the main, headquartered in the United States. All of the Five are both headquartered and incorporated in the United States, and their lobbying efforts in Washington, D.C., as well as in state legislatures, are quite sophisticated. Indeed, the tight control tech holds over Congress is one of the reasons an agreement on a global tax rate for tech is in the works in the first place. That world leaders are looking for global answers to the problems with regulating tech is a reason to be optimistic about our technological future. No matter where a particular company is head-quartered, tech is by its nature a global industry without boundaries, available wherever a consumer can access a broadband signal. Practically speaking, then, if the E.U. imposes regulations on the industry, those regulations mean less than nothing in the United States, India, and China. Conversely, if the United States creates regulation, those rules and laws will have no impact on tech's operations in any other country, either. The only way to break the hold tech has over governments all over the world is for the world to come together and create consistent boundaries for tech to function within.

Take a step back with me to understand the scope of the tax problem we're facing in regard to Big Tech. In June of 2021, ProPublica "obtained a vast trove of Internal Revenue Service data on the tax returns of thousands of the nation's wealthiest people, covering more than 15 years."[446] This material provided an unprecedented, inside look at a tax system rigged to benefit the very wealthiest among us. "To take one example, [Jeff] Bezos's wealth soared by $120 billion from 2006 to 2018, and his federal taxes during that time amounted to only 1.09 percent of the wealth gain. The situation for the average household was radically different: Its taxes amounted to more than 100 percent of its wealth increase."[447]

Now, this isn't a book about taxes, but the fact that the twenty-five richest individuals in the country have been proven to be so successful in using the existing tax system to avoid paying their taxes is hard proof of something we already knew: it is the system that is unfair. The way we tax both individuals and corporations is part of our regulatory framework that does not work, and it needs to work because countries pay their bills with the taxes they collect. They build roads and schools, insure banks

---

[446] Jesse Eisinger, Jeff Ernsthausen and Paul Kiel, "The Secret IRS Files: Trove of Never-Before-Seen Records Reveal How the Wealthiest Avoid the Income Tax," *ProPublica* (June 8), https://www.propublica.org/article/the-secret-irs-files-trove-of-never-before-seen-records-reveal-how-the-wealthiest-avoid-income-tax.

[447] David Leonhardt, "A Voluntary Tax," *New York Times* (June 9, 2021), https://www.nytimes.com/2021/06/09/briefing/tax-jeff-bezos-wealthy.html.

and maintain their armed forces, prosecute crime and support the arts, inspect meat packing plants and water filtration systems, fund police forces and post offices and public parks and museums. Corporations, as well as people, have to pay their fair share to fund the public services that they and their employees use in order for a country to operate in a way that benefits the people who call it home. An across-the-board, global tax rate closes the loophole of tax havens to which they could escape if the United States alone raised taxes on tech, and it solves one part of the problem.

The other part of the problem is that countries that don't serve as headquarters for a tech company don't collect income taxes on that company, even though their citizens use the company's services as robustly as Americans do. Amazon, Google, Microsoft, Facebook, and Apple consistently rank at the top of the list of the world's most profitable companies. So, countries around the globe need to have an instrument through which to collect a share of tech profits commensurate with the way their citizens use tech.

Because tech *isn't* headquartered in those other countries—and because, in consequence, tech doesn't lobby their governments with the same fervor it does in the United States—other countries can more easily take action to regulate tech than we can. The E.U., for one, saw an opening, and they took it. Back in 2016, they enacted the General Data Protection Regulation (GDPR),[448] which regulates data protection and privacy in the E.U. and the European Economic Area (EEA). It includes an outline of rights of the data user, particularly concerning transparency and the right to access their own information; principles under which data can be lawfully collected from an individual; and penalties for companies that don't comply. The GDPR is being used as a model for other countries who are in the process of creating their own data protection regulations, from Japan to Argentina to Kenya, and the California Consumer Privacy Act (CCPA), enacted in 2018, shares many of its provisions with the GDPR. Indeed, through enactment of the GDPR,

---

[448] GDPR website, https://www.gdpreu.org.

the E.U. has rather forced these other countries, including the United States, into adopting the parameters it has already defined. Coming in so far behind the curve, "legislators in Washington [are left with] no choice but to use those laws as de facto standards."[449]

The governing bodies of the E.U. had an easier time keeping pace with the innovators, and as I've noted, this is in part because those innovators *weren't* headquartered in their countries. The E.U. was able to look at what was happening from a distance, seeing a larger part of the forest because they were not directly in the midst of the trees.

o o o

Among the other reasons the E.U. has pulled so far ahead of the United States in dealing with Big Tech and its ramifications for its citizenry is that in Europe there is a broader cultural acceptance of government regulation. In the United States, we have "rugged individualism" and unfettered capitalism; in Europe, the economy operates in a much less *laissez-faire*—French for "allow to do"—manner. There is more general agreement that government oversight is both natural and beneficial.

In the E.U., the top tech regulator is Margrethe Vestager,[450] Executive Vice-President of the European Commission, who chairs the Commissioners' Group on a Europe Fit for the Digital Age. Her goal in this role "presents a European society powered by digital solutions that put people first and simultaneously open up new opportunities for businesses. Furthermore, this initiative will boost the development of trustworthy technology to foster an open and democratic society and a vibrant and sustainable economy."[451] The United States has, at the time

---

[449] Ina Fried, "How the U.S. got boxed in on privacy," AXIOS (June 9, 2021), https://www.axios.com/newsletters/axios-login-d573aea0-1c37-429e-812b-5a430f23bb2b.html.

[450] "College (2019–2024) The Commissioners," *European Commission,* https://ec.europa.eu/commission/commissioners/2019-2024_en.

[451] "A Europe Fit for the Digital Age," *Open Access Government* (June 25, 2020), https://www.openaccessgovernment.org/a-europe-fit-for-the-digital-age/89167/.

of this writing, no comparable counterpart to Ms. Vestager at work on the same sort of "human-centric"[452] goals for our technological future.

Working with Ms. Vestager is Vera Jourova, European Vice President for Values and Transparency.[453] Among her duties are tabling legislative proposals to ensure more transparency in paid political advertising and clearer rules on financing European political parties, and supporting work on countering disinformation and fake information, while preserving freedom of expression, freedom of the press, and media pluralism. In other words, she is the E.U.'s top disinformation official—again, a role for which there is not yet a counterpart in the United States.

When Ms. Vestager spoke with Kara Swisher in June 2021 on the podcast *Sway*, she answered Ms. Swisher's question regarding the effect of regulation on corporate growth and success in this way: "...the only thing that regulation and competition law enforcement stifles is innovative attempts to break that regulation and antitrust law enforcement.... it's about time that democracy catches up and gets a bit ahead of technology because it's not by our publicly elected representatives that our society is being shaped. It is by corporate business. But it is 100% legitimate for our elected representatives to say 'this is the framework within which you have to go out and compete'. And, yes, that puts a brake on some things, but then it's democratically decided. And I think that is perfectly fine."[454]

This is the sort of pragmatic, clear-eyed point of view that can result when there are tech-literate people in dedicated positions of real authority who are not the subject of constant and well-funded lobbying efforts by special interests. The United States desperately needs counterparts to Ms. Vestager and Ms. Jourova in order to move ahead and both efficiently and beneficially sort out our relationship to tech.

---

[452] Ibid.

[453] "Vice-President (2019–2024) Vera Jourova," *European Commission*, https://ec.europa.eu/commission/commissioners/2019-2024/jourova_en.

[454] "Meet Big Tech's Tormenter in Chief," *New York Times* (June 10, 2021), https://www.nytimes.com/2021/06/10/opinion/sway-kara-swisher-margrethe-vestager.html?showTranscript=1.

o o o

Yes, the United States is running behind our European counterparts when it comes to tech. But I don't mean to imply we can't catch up. The breakthrough agreement about a global tax rate for tech at the June 2021 G-7 is evidence that our leaders are, at last, taking a more proactive stance in tech regulation. It's a gateway toward creating the landscape we need to create a workable formula for dispensing data dividends. As always, though, it will be our activism, the pressure from the people those leaders govern, that will crystalize and sustain the urgency around the issue.

So, what does this activism look like? What can we do for ourselves, in our homes, to make our personal devices safer for ourselves and our families—and what can we do on the larger stage to encourage U.S. leaders and regulators to not only continue their efforts but to ramp them up?

The first, large-stage consideration must be to remove the shield Big Tech has used for so long to deflect any liability for the disinformation that runs rampant on its platforms.

Section 230 is part of the Communications Decency Act, a 1996 law which is itself part of the Telecommunications Act of the same year that regulated online pornography. Section 230 provides, additionally and specifically, for legal immunity from liability for internet services and users for content posted on the internet, with some exceptions regarding copyright infringement, and federal crimes, such as child sex trafficking—though we cannot go on without acknowledging that a full 59 percent of sex trafficking victims in the United States in 2020 were recruited on Facebook.[455]

Specifically, the protection to Big Tech under the current version of Section 230 is: "No provider or user of an interactive computer service shall be treated as the publisher or speaker of any information speech provided by another information content provider."

---

[455] Chris Jewers, "More Than Half of Sex Trafficking Victims in the U.S. Last Year Were Recruited on Facebook, Bombshell Report Reveals," *Daily Mail* (June 10, 2021), https://www.dailymail.co.uk/news/article-9672795/Over-half-sex-trafficking-victims-U-S-year-recruited-Facebook-report-reveals.html.

But, wait a minute—I can almost hear you thinking. In 1996, what did internet service providers need immunity *from*?

In 1996, two Stanford University students, Larry Page and Sergey Brin, were still developing a search algorithm they called "BackRub"; Google.com wasn't registered until 1997 and not incorporated until 1998. In 1996, Microsoft, which was founded in 1975—what seems, in this context, like pre-history—was just a year into the launch of its first web browser that came packaged in the Windows 95 Plus! Pack, but not in Windows 95, because the boxes were printed before the web browser was finished. In 1996, Amazon was just one year old and wouldn't make its initial public offering—at eighteen dollars a share—until 1997. In 1996, Apple, which had been established in 1977, was in a hot financial mess with low cash liquidity, low quality products, no viable strategy for an operating system, and still a year away from bringing back co-founder Steve Jobs—as a consultant. In 1996, Mark Zuckerberg was living in Dobbs Ferry, New York, where he was an exemplary sixth-grade student.

Well, travel with me a little farther back in time.

In 1959, the Supreme Court heard *Smith v. California*, and its ruling in this case drew a clear line between the publisher of content and the distributor of that content. It established that a publisher of content—say, a book or a magazine or a newscast—would certainly have knowledge of the material it was putting out into the public, so it could be held liable for illegal content; but a distributor—say, a station that aired a newscast, or a newsstand that sold magazines, or a bookstore that sold books—would not necessarily have knowledge of the material contained in its wares, and so could not be held liable.

Jump forward to the early 1990s, when the internet was just beginning to be adopted by users, and a number of legal challenges were brought against service providers—notably Prodigy and CompuServe—for content that had been generated by its users. Because Prodigy had employed a team of moderators to fact-check content and remove illegal posts, the Court determined that it had taken an editorial role in the material it was putting out into the public, and the company was held liable for illegal material. CompuServe, on the other hand, had adopted a policy of not moderating its content, so it was found to be a mere distributor, and the company was not held liable for illegal content.

The rulings were not met with universal approval. If a company, a service provider, or platform was to be held responsible for any illegal thing any one of the growing millions of users might post, then that was going to create a risk of exposure that most companies would not be able to absorb—and it would severely limit the growth of the then-nascent internet. If, on the other hand, providers and platforms, under threat of being held responsible for whatever any user or users might post, declined to moderate their users' content, then the internet was likely going to become a wildly uncivil and even incendiary place.

Then-Representatives Christopher Cox (R-CA; now retired) and Ron Wyden (D-OR; now Senator Wyden) attempted to solve the problem by creating Section 230 as a part of the Communications Decency Act then being debated in Congress. Since 1996, section 230 has guided the assignment of liability for online crimes and torts by content creators, platforms, and service providers. Fundamentally, the statute's main principle is that content creators should be responsible for any illegal content they create. But internet platforms and providers are, generally, exempt from liability for third-party—meaning user—content. The caveat here is that if the platform or provider creates the illegal content, is aware of hosting illegal content, or violates federal criminal laws in any other way, then Section 230 offers them no protection.[456]

Why an internet service should be thus protected is, in one important way, understandable. The content of Facebook or Twitter is almost entirely user-generated, and policing whatever a world of users will post on their individual pages and feeds is, at best, unwieldy—and grows ever more so the larger the user base grows. And, for a variety of reasons, we *do* want the user base to grow. After all, these are social *networks* we're talking about, and, by definition, the larger a network, the more useful it is. But just because a problem is large and cumbersome doesn't mean

---

[456] "Content Moderation: Section 230 of The Communications Decency Act," *Internet Association*, https://internetassociation.org/positions/content-moderation/section-230-communications-decency-act/#.

we get to either ignore it or abdicate responsibility for it. Lawmakers who are tempted to do so—for example, decrying such small regulation as Section 230 currently imposes, and opposing reforms for the purpose of broadening and hardening its scope—need to be reminded, quickly and often, of Facebook's sex trafficking problem. A problem, we should note, that got worse in June 2021, when the Texas Supreme Court ruled that Facebook can be held liable for sex trafficking on its platform, in spite of the protections under Section 230. In their majority opinion, the justices wrote that Section 230 does not "create a lawless no-man's-land on the internet."[457]

We can call this a good start even while acknowledging that the problem of how to fix Section 230 truly is a thorny one. Still, pause here for a moment to marvel at how far behind the United States truly is in terms of keeping up with the world of technology. In this fast-paced, innovative atmosphere, where a user's relationship with technology changes materially even from year to year—and where simply anecdotal evidence proves that the internet has indeed become an uncivil and incendiary place—the United States is still functioning under a statute that was enacted into law a quarter century ago.

There is significant disinformation swirling around about Section 230 itself. The lawmakers who denounce the thin layer of regulation that Section 230 currently imposes, calling it *censorship*, really don't have a good grasp of what they're talking about. As it stands, Section 230 protects Big Tech from liability when a user or users post or tweet information that is false, misleading, or designed to incite extreme reaction. The consequence is that users can post and tweet without concern that Big Tech will, for example, fact-check them—and Big Tech can continue to employ lax content policies. However, should this protection from liability be removed through the repeal of Section 230, and Big Tech

---

[457] Katherine Huggins, "Court Rules Facebook Can Be Liable for Sex Trafficking on Its Platform: Section 230 Does Not 'Create a Lawless No Man's Land,'" *Mediate* (June 26, 2021), https://www.mediaite.com/online/court-rules-facebook-can-be-liable-for-sex-trafficking-on-its-platform-section-230-does-not-create-a-lawless-no-mans-land/.

become accountable for publishing false or inflammatory content on its platforms, that's when Big Tech will become truly interested in things like fact-checking—and that could likely be the starting point for a real problem with censorship.

The question then becomes: does Big Tech have the *right* to do things like fact-check and, potentially, delete a user's post because it does not measure up to standards that might be set around truthfulness or decency? Often, when this question is posed, the answer will invoke the text of the First Amendment in the Bill of Rights, and an argument will be made that any attempt by Big Tech to monitor user content is an infringement on the user's right to free speech.

But let's look at the actual text of the First Amendment: "Congress shall make no law respecting an establishment of religion, or prohibiting the free exercise thereof; or abridging the freedom of speech, or of the press; or the right of the people peaceably to assemble, and to petition the Government for a redress of grievances." Let's be very clear about this: the operative words in that text are "Congress shall make no law." This means that the First Amendment guarantees that the *government* shall make no law prohibiting free expression. Public companies—like Facebook and Twitter and YouTube and Google and all the rest—are not now and have never been bound by the First Amendment. If an owner of a movie theatre has an employee who stands in the middle of his crowded theatre and yells "Fire!", causing panic among the moviegoers, then the movie theatre owner is perfectly within his rights to terminate that employee. Likewise, if the owner of an internet platform has a user who yells "The election was stolen!", causing panic among the electorate, then the platform owners are perfectly within their rights to block that user's account.

For clarity, let's dive a little deeper into First Amendment history and law. In 1969, the Supreme Court of the United States handed down its landmark decision in the case *Brandenburg v. Ohio*.[458] In this case, Clarence Brandenburg, a Ku Klux Klan leader in rural Ohio, was, in 1964, charged with and convicted of advocating violence under Ohio's

---

[458] "Bradenburg v. Ohio, 395 U.S. 444 (1969)," *Justia*, https://supreme.justia.com/cases/federal/us/395/444/.

criminal syndicalism statute for his involvement in a rally during which a cross was burned, speeches were made urging the forced expulsion of African Americans and Jews from the United States, and accusing the president, Congress, and Supreme Court of colluding with non-whites against whites. The Supreme Court, however, overturned Brandenberg's conviction, arguing that certain conditions must be met in order for criminal liability to be imposed in this case. Does the speech incite others to illegal actions or imminent harm? Is there a likelihood that the incited illegal action will take place? And, was it the intent of the speaker to cause such illegal action to take place? All these years later—and though it has come under scrutiny in the face of the rising threat of global and domestic terrorism—*Brandenberg v. Ohio* remains the Supreme Court's last word on free speech, and the precedent remains the principal standard in deciding cases in this area of First Amendment law.

But, again, Big Tech—Facebook and Twitter and Google and all the rest—are public corporations. Though at this point in our history they may wield the power of stand-alone nations, they are *not* government, and therefore they have every right to monitor the content their users post online, and to remove it if and as they see fit. *Governments* are the entities that have no rights in terms of monitoring the speech of their citizens. For instance, Florida's Senate Bill 7072—which calls for hefty fines against any social media company that de-platforms a candidate for political office, and which was signed into law in May 2021 by Governor Ron DeSantis (R-FL), a likely future presidential candidate—is ripe for challenge in court. "The First Amendment prohibits the government from compelling or controlling speech on private websites," according to Carl Szabo, the president and general counsel at Net Choice. "By forcing websites to host speech, this bill takes us closer to a state-run internet where the government can cherry pick winners and losers."[459] This same reasoning is why most experts are predicting that the lawsuit brought

---

[459] Catherine Thorbecke, "Critics Slam Florida's Law Banning Big Tech 'De-Platforming' As 'Unconstitutional,' *ABC* (March 25, 2021), https://abcnews. go.com/Technology/critics-slam-floridas-law-banning-big-tech-de/story?id= 77891650.

by Donald Trump in the summer of 2021, suing certain social media platforms for ostensibly infringing on his right to free speech, will not be successful.[460]

Perhaps, in this light, it is more appropriate to ask if Big Tech has, not the right, but the *responsibility* to do things such as fact-checking user content. Mike McCue, co-founder and CEO of Flipboard, makes the case that applying journalistic standards and principles to news and other information delivered on social media platforms could be a large part of curbing the flood of misinformation and disinformation in which social media is now awash, and helping to restore consumer trust in the quality of information they receive. "One of the risks for the tech industry that values what's new, who's hot, and what's next is a failure to draw on knowledge gained through experience and the people who hold the keys to that knowledge. If a company has the potential to disrupt, then it must understand the industry it could purposefully or accidentally upend. Tech platforms failed to do this with journalism, and we as a society are paying the price."[461]

The adoption and rigorous application of journalistic standards by social media content creators, let alone more extreme partisan platforms, might work in a more perfect—or at least less riven—world, but as a voluntary undertaking, it would probably not work in our contemporary reality. So what are the options now in play for refining the law around who bears responsibility for illegal content?

There is a school of thought that the Federal Communications Commission (FCC) has the authority to interpret Section 230, and therefore to construct the legal framework through which the law is applied. Indeed, this wording appears on the FCC's web page: "But in

---

[460] Fred Hiatt, "Legally, Trump's tech lawsuit is a joke. But it raises a serious question.," *Washington Post* (July 8, 2021), https://www.washingtonpost.com/opinions/legally-trumps-tech-lawsuit-is-a-joke-but-it-raises-a-serious-question/2021/07/08/33bc2dfa-e010-11eb-9f54-7eee10b5fcd2_story.html.

[461] Gabriella Schwarz, "Why Tech Platforms Need to Be Built on Journalistic Values," *Nieman* (February 20, 2019), https://nieman.harvard.edu/articles/why-tech-platforms-need-to-be-built-on-journalistic-values/.

my own judgment, the FCC's legal authority to interpret Section 230 is straightforward: Congress gave the Commission power to interpret all provisions of the Communications Act of 1934—including amendments—and Section 230 is an amendment to the Communications Act. The Commission therefore may proceed with a rulemaking to clarify the scope of the Section 230(c) immunity shield."[462] This opinion, written in October 2020 by Thomas M. Johnson, Jr., FCC General Counsel, goes on to say that the agency's role as "'authoritative interpreter' may be particularly useful where…courts have reached divergent interpretations of key provisions of an important statute, thus creating substantial uncertainty and disharmony in the law."[463]

This, however, was an opinion offered in 2020, during the Trump years, when the administration was attempting to use Section 230 to "prevent websites from moderating content in a way that many conservatives believe is biased against them."[464] It's important to note that the FCC is under the direction of five commissioners who are appointed by the president, and confirmed by the Senate for five-year terms. No more than three of the commissioners may be from the same political party, and yet within the arrangements for the commissioners' appointments is an inherent vulnerability to manipulation for political ends. Therefore, interpreting Section 230, and giving that interpretation the authority of law, needs to fall to the legislative branch.

There are several paths Congress has been debating about the way forward on Section 230, and most of them circle around algorithms. Senator Ben Sasse (R-NE) and another probable future presidential candidate, has said, "'I think it's very important for us to push back on the

---

[462] Thomas M. Johnson Jr., "The FCC's Authority to Interpret Section 230 of the Communications Act," *Federal Communications Commission* (October 21, 2020), https://www.fcc.gov/news-events/blog/2020/10/21/fccs-authority-interpret-section-230-communications-act.

[463] Ibid.

[464] Sara Morrison, "What the FCC can and can't do to Section 230," *Vox* (October 21, 2020), https://www.vox.com/recode/21519337/section-230-trump-fcc-twitter-facebook-social-media-ajit-pai.

idea that really complicated, qualitative problems have easy quantitative solutions,' [and he] argued that because social media companies make money by keeping users hooked to their products, any real solution would have to upend that business model altogether."[465] Yes; it would. And there are four proposals now on the table that would do just that.

The first is the "Don't Push My Button" bill, sponsored in the House by Representative Paul Gosar (R-AZ) and Tulsi Gabbard (D-HI). This bill would deny Section 230 immunity to any platform that uses "algorithms to optimize engagement by funneling information to users that polarizes their views, *unless a user opts into such an algorithm.*"[466] (Emphasis mine.) What this means is that those who subscribe to extremist views could still access the full body of on-going misinformation and disinformation, as long as they told Facebook or Twitter or new-as-of-summer-2021 Trump-affiliated social media experiment GETTR[467] that they wanted to receive it. This particular bill wouldn't stem the flow of misinformation and disinformation, only more efficiently funnel it to an already extremist base.

The second bill is titled the "Protecting Americans from Dangerous Algorithms Act," which was reintroduced in Congress in March 2021 by Congresswoman Anna G. Eshoo (D-CA) and Congressman Tom Malinowski (D-NJ). This bill "narrowly amends Section 230 of the Communications Decency Act to remove liability immunity for a platform if its algorithm is used to amplify or recommend content directly relevant to a case involving interference with civil rights (42 U.S.C. 1985); neglect to prevent interference with civil rights (42 U.S.C. 1986); and in

---

[465] Taylor Hatmaker, "At Social Media Hearing, Lawmakers Circle Algorithm-Focused Section 230 Reform," *TechCrunch* April 27, 2021), https://techcrunch.com/2021/04/27/section-230-bills-algorithms-congress-hearing/?guccounter=1.

[466] John Kennedy website, https://www.kennedy.senate.gov/public/2020/10/kennedy-applauds-house-companion-to-don-t-push-my-buttons-act.

[467] Grace Panetta, "Trump's former top aide launches GETTR, a new conservative social media platform," *Business Insider* (July 1, 2021), https://www.businessinsider.com/gettr-trump-new-social-media-platform-2021-7.

cases involving acts of international terrorism (18 U.S.C. 2333)."[468] The "Protecting Americans from Dangerous Algorithms Act" is somewhat stronger than the "Don't Push My Button" proposal, in that it would attempt to completely curtail the spread of some of the most alarming types of misinformation and disinformation, rather than allowing them to flourish for a self-selected audience.

The third bill under consideration is one proposed by Senators Amy Klobuchar (D-MN), Mazie Hirono (D-HI), and Mark Warner (D-VA), called the "SAFE TECH Act." Under this legislation, the change to Section 230 would be disarmingly subtle: the language of the statute would stay the same, as would the protection from liability *except in cases where payment was involved.* Specifically, the proposed change to the language of Section 230 is: "No provider or user of an interactive computer service shall be treated as the publisher or speaker of any speech provided by another information content provider, except to the extent the provider or user has accepted payment to make the speech available or, in whole or in part, created or funded the creation of the speech."[469] The idea behind this suggested change is, as Senator Warner puts it, online ads that are "a key vector for all manner of frauds and scams,"[470] so the goal is to zero in on abuses to the system as it is currently in place. Then, however, the problem becomes to define exactly what constitutes "payment." Obviously, this would include straightforward payment for ad buys of the sort sellers and PACs and the like make on platforms such as Facebook, Twitter, Google, and Amazon every day. But would it also include payments made to content creators on platforms such as Patreon[471]

---

[468] "Reps Eshoo and Malinowski Reintroduce Bill to Hold Tech Platforms Accountable for Algorithmic Promotion of Extremism," Eshoo website (March 24, 2021), https://eshoo.house.gov/media/press-releases/reps-eshoo-and-malinowski-reintroduce-bill-hold-tech-platforms-accountable.

[469] Taylor Hatmaker, "The SAFE TECH Act Offers Section 230 Reform, But the Law's Defenders Warn of Major Side Effects," *TechCrunch* (February 5, 2021), https://techcrunch.com/2021/02/05/safe-tech-act-section-230-warner/.

[470] Ibid.

[471] Patreon website, https://www.patreon.com.

and Medium?[472] Could the interpretation ultimately be stretched to include situations where payment was made for an ad's design to a third-party graphic artist? Would it include the as-yet-undefined dollar value of the payment users make to Big Tech every hour of every day, in the form of offering up their private information for use by and profit to the largest corporations? As Senator Ron Wyden put it, "Creating liability for all commercial relationships would cause web hosts, cloud storage providers and even paid email services to purge their networks of any controversial speech."[473]

Finally, there is Representative David Cicilline's (D-RI) proposed legislation, which would also focus on impacting and refining the decisions Big Tech makes around how they use algorithms to amplify user engagement. Representative Cicilline runs the House Judiciary Committee's antitrust panel, and he has a strategy apart from his colleagues. Rather than "give the major technology companies and their armies of lobbyists the easy target of a massive antitrust bill,"[474] his plan is to create a series of smaller bills to make it "harder for Amazon, Facebook, Apple and Google to mobilize quickly against reforms they don't like."[475] In addition to presenting Big Tech with a moving target, of sorts, Representative Cicilline says that his plan for a series of smaller bills might have a better chance at bipartisan support as legislators would be able to attach themselves to certain areas of reform that they like, and not to others that they don't.

While this strategy can seem workable on its surface, concern arises around unforeseen, and unintended, consequences of addressing the reformation of Big Tech in a piecemeal rather than a holistic way. That is, if we return to Senator Sasse's thought on the subject—that any effective solution will involve upending the flawed business model under

---

[472] Medium website, https://medium.com.

[473] Hatmaker, ibid.

[474] Jonathan Swan, "Inside the Democrats' strategy to bombard Big Tech," AXIOS (March 21, 2021), https://www.axios.com/tech-antitrust-facebook-google-amazon-apple-275f122d-b3f5-49cb-b223-f77c95a49252.html.

[475] Ibid.

which tech now functions—it would behoove us, as well as Big Tech, to craft a solution that encompasses a healthy and well-rounded vision of what Big Tech could be, going forward. Think again of how I described algorithms—as a set of instructions or *recipes* designed to create a specific outcome. What would happen if we gave ten different cooks one recipe to make a lasagna, for example, and asked them each to prepare one of the ingredients for the dish, but we also told them that each ingredient was only a suggestion? That they could support the use of their ingredient or not, as they saw fit? The cook who was assigned to make the tomato sauce might be allergic to garlic, so he leaves out that ingredient as he simmers his sauce. The cook who was tasked with preparing the ricotta layer—mixing it with eggs and mozzarella and parsley—might not like ricotta and decide he'd rather make a dessert, so he mixes up some mascarpone with honey and lemon. Another of the cooks might be on a low-carb diet and decide that the dish can forego noodles altogether. I don't know what you'd end up with at the end of the experiment, but it sure wouldn't be a lasagna.

Another way to approach the refinement of Section 230 would be to start treating digital publishers in the same way that broadcasters are treated. That is, radio and television broadcast stations are regulated by the Federal Communications Commission (FCC), which issues each station a license to air its programming. And "in exchange for obtaining a valuable license to operate a broadcast station using the public airwaves, each radio and television licensee is required by law to operate its station in the 'public interest, convenience and necessity.'"[476]

At the outset, this idea can seem self-evident, but it brings up several sticky points. First, there's the little matter of who owns the method of transmission. The airwaves have long been deemed to belong to the public, but there are a lot of entities—Internet Service Providers (ISPs), Local Area Networks (LANs), Internet Exchange Points (IXPs), as examples—many of them private, that are heavily invested in their ownership of a piece of the net. Beyond this rather basic issue, we run into

---

[476] "The Public and Broadcasting," *Federal Communications Commission* (Revised September 2021), https://www.fcc.gov/media/radio/public-and-broadcasting.

a problem of interpretation, similar to the issue we have with creating a hard-and-fast, one-size-fits-all interpretation of the First Amendment: who is going to be empowered to determine what is "in the public interest, convenience and necessity"? As we've discussed previously, the FCC is run by five presidential appointees, and so is vulnerable to exploitation for political purposes by an unethical administration. Therefore, the task of determining what the parameters are for an internet service to act in the public interest necessarily falls to the legislature. Given these concerns, it might be that revisiting the Fairness Doctrine and translating it into a form that could be applied to the internet and internet platforms is worth consideration.

The Fairness Doctrine, which was introduced in 1949, was a policy of the FCC that required those who held a broadcast license to present both sides of a controversial issue, and to do it in a way that was equitable and honest. The Doctrine differed from the equal-time rule, which is still in effect, in that providing equal time applies only to political candidates and not to issues under debate by the general public. Contrasting points of view could be presented in a wide variety of formats—through editorials, public-affairs shows, news segments—as long as both sides got a fair hearing on the public airwaves.

In 1985, however, Mark S. Fowler, FCC Chairman under then-President Ronald Reagan, began the ultimately successful attempt to repeal the policy. The Fairness Doctrine was struck down in 1987, but not without a fight. Before the policy was done away with, Congress tried to codify the Doctrine into law, but that legislation was vetoed by Reagan. In 1991, an attempt was made to revive the Doctrine, but then-President George H. W. Bush threatened to veto it even before it could get to his desk. There is a reason to believe that the demise of the Fairness Doctrine has played a part in our country's contemporary political polarization. Can we consider it a coincidence that it was in 1987 that Rush Limbaugh—then an employee at a small radio station in Sacramento, California, who "[f]rom his earliest days on the air...trafficked in conspiracy theories, divisiveness, even viciousness"[477]—was signed by ABC Radio, and conservative talk radio was born?

But what would a digital Fairness Doctrine look like? To whom would it apply? Would platforms like Facebook and Google be required to present contrasting points of view in newsfeeds and search results? Certainly, it couldn't fall to the millions of social media users themselves to present both sides of an argument in every post they created to express their own opinion on an issue. How, in that case, would—or could—it be enforced? Wouldn't that be its own infringement on the users' right to free speech?

As you can see, the form that Section 230 will take in the future is a tricky issue. But the events of January 6, 2021, make it clear we can't continue to function peaceably as a nation under the lax social media content policies as they now exist. In order to avoid a scenario in which we're dealing after the *next* crisis with various sets of confusing—and, perhaps, conflicting and even crippling—social media policies, set hurriedly, arbitrarily, and piecemeal by each individual social media entity or state in the union, it's imperative that our lawmakers take the matter into hand now, and handle it as thoughtfully and seriously as it deserves.

⌒ ⌒ ⌒

In addition to reforming Section 230, there are other ways in which we can—and should—handle the burgeoning issues around Big Tech. Fortunately, these additional measures aren't quite as fraught as finding and agreeing upon a definition of free speech.

The urgency with which we must address the first of these measures was thrown into stark relief during the pandemic of 2020–21. What we learned from the more than year we spent, for the most part, locked in our homes and isolated from most forms of usual communal activities, is how critical technology is to both our material and our psychological

---

[477] Paul Farhi, "Rush Limbaugh is ailing. And so is the conservative talk-radio industry.", *Washington Post* (February 9, 2021), https://www.washingtonpost.com/lifestyle/media/rush-limbaugh-conservative-talk-radio/2021/02/09/97e03fd0-6264-11eb-9061-07abcc1f9229_story.html.

well-being. For fifteen months of our lives, many of us worked from home, communicating with our offices through email and taking meetings on Zoom. Our children attended school online, many of them passing from one grade to another without ever entering a classroom. Our families attended worship services at churches and synagogues and mosques over the internet. We kept in touch with friends and family by way of Facetime. Brick-and-mortar stores were closed, and we couldn't leave our homes to go to them, anyway, so we ordered in groceries and meal kits and take-out, clothing and gifts, medications and cosmetics— and that's just the short list. We depended on cable channels to bring us news of the outside world, and when vaccines were at last available to help bring an end to the crisis, we searched on our computers or mobile phones to secure appointments to get our jabs. We couldn't go to stadiums to see ball games or concerts in 2020, but celebrities like John Legend, Pink, and Keith Urban streamed concerts to keep up the spirits of all of us who were homebound. Indeed, we relied nearly exclusively on technology to entertain us. At the height of the lockdown, adults were spending an average of nearly six-and-a-half hours a day in front of a screen, time spent on subscription streaming services doubled, and twelve million people joined a service they had never used pre-pandemic.[478]

Twitch,[479] for example, which is owned by Amazon, is a video-streaming service that began in 2011 and offers a wide range of topics to their viewers, from cooking to music. But its main traffic driver is video games—that is, to be clear, not its users actually playing video games, but watching other people play video games. An audience that is comprised of about 66 percent men, generally between the ages of eighteen to thirty-four, spends an average of twenty-nine minutes *three times a day* watching other people play video games. And who are those people playing the games? That's another part of the Twitch phenomenon. If the internet has democratized information distribution—and it has—then Twitch is the proof that anyone can be a content creator. They even have

---

[478] "TV Watching and Online Streaming Surge During Lockdown," *BBC* (August 5, 2020), https://www.bbc.com/news/entertainment-arts-53637305.

[479] Twitch TV website, https://www.twitch.tv.

a "Twitch Creator Camp" to teach their users how to create and monetize their content.[480] The platform has successfully turned gaming into a spectator sport, and it is now the most watched streaming TV service behind platforms like Hulu and Netflix. Twitch sees more than fifteen million daily active users and has over 3.8 million unique broadcasters,[481] according to published reports, though a senior industry insider has informed me that, through the pandemic, Twitch saw almost *thirty million* users every day.

In short, if we hadn't realized it before a virus confined us to our homes, technology is an integral part of our lives, and it was likely essential to our survival, or at least our sanity, in 2020.

The hard truth is that technology made negotiating the pandemic not only easier for most of us, but—thanks to essential workers who kept the supply chains going throughout the crisis—likely saved countless lives by allowing so many of us to stay home, where we remained many degrees removed from transmission opportunities. It was as elemental to our comfort and endurance through the hardship as was running water and electricity. This fact renewed many people's interest in the idea of broadband as a utility—just as running water and electricity have become utilities as they grew more common over the decades and were deemed essential services.

A utility is a company that supplies a community with water, sewerage, electricity, or gas, but it doesn't simply deliver the end-product, or service, to your home or business. It is also responsible for the maintenance of the lines, poles, pipes, and meters it requires to do its work. If a storm causes a power outage, or a truck hits a utility pole, the company is responsible for the repairs and for restoring your service.

Internet access as a utility isn't a new idea. Through the 1990s and into the early 2000s—when dial-up access was painfully slow and required a dedicated phone line, and when many less people relied on it for either work or pleasure—the internet was a luxury. But then broadband

---

[480] Ibid.

[481] Joseph Yaden, "What is Twitch?" *Digital Trends* (March 15, 2021), https://www. digitaltrends.com/gaming/what-is-twitch/.

was introduced. At first, this was a luxury item of sorts, as well. The cost to the consumer was high but, as more Internet Service Providers (ISPs) entered the game and competition increased, the cost decreased and broadband became ever-more affordable. More people began to rely on broadband for their home computers, as did more businesses, as did more hotels and vacation resorts, until today you'd be hard-pressed to find a coffee shop without a connection.

That is, except in rural neighborhoods, and small towns, and certain areas of cities where there isn't a connection, or at least a reliable one. The "digital divide" has been a concern among policymakers for at least a decade. It refers to a socioeconomic distinction: a separation of those who live in middle-class and upper-middle-class places, where broadband exists and can be accessed, from those who live in poorer areas from which providers can't project profitable returns and therefore don't invest. The pandemic exponentially magnified the extent to which this divide impacts the lives of those who lack access—from schoolchildren who were unable to logon to attend classes remotely, to families facing unemployment but were unable to access portals to register for benefits. These days, it isn't only politicians who align with Bernie Sanders who advocate for broadband as a utility[482]—Justice Clarence Thomas has proposed that companies like Facebook, Twitter, and Google should be regulated as utilities, too.[483] The political philosophy from whence the position is derived might come from completely opposite sides of the spectrum, but it is on the radar of both—and Big Tech is unnerved by the rumblings.

Even more unsettling to the tech industry is the antitrust fervor rising across the globe. *Antitrust* is a term that refers to legislation designed to prevent or manage monopolies in the interest of promoting competition.

---

[482] Ben Gilbert, "Bernie Sanders Has a $150 Billion Plan to Turn the Internet into a Public Utility with Low Prices and Fast Speeds––Here's How His Plan Works," *Insider* (January 22, 2020), https://www.businessinsider.com/bernie-sanders-internet-as-utility-plan-explainer-2019-12.

[483] Issie Lapowsky, "Justice Thomas argues for making Facebook, Twitter and Google utilities," Protocol (April 5, 2021), https://www.protocol.com/bulletins/thomas-scotus-twitter-trump-ban.

In America, Senator Amy Klobuchar is one of the most prominent faces of the movement. She's chair of the Senate Judiciary Committee's antitrust subcommittee and author of *Antitrust: Taking on Monopoly Power from the Gilded Age to the Digital Age*, in which she makes "the historical and present-day argument that one way to remedy the growing monopolization in our country is to actually do something to counteract consolidated business power."

The historical perspective Klobuchar is talking about is, of course, the Gilded Age, the period of time between the Civil War and the turn of the twentieth century: the apex of the Industrial Revolution, and the era of the original "robber barons"—Rockefeller, Frick, Carnegie. But, as I've mentioned earlier in this book, it was known as well for its more sinister side: the wealth and greed of the industrialists, bankers, and politicians who lived opulently at the expense of the working class—and the muckrakers like Lincoln Steffens, Upton Sinclair, and Ida Tarbell who worked tirelessly to expose the corruption of the tycoons and bust the monopolies they held in railroads, oil, and steel. What is less known about this era is that the board game "Monopoly" was invented at this time—by a stenographer named Elizabeth Magie, for the purpose of educating the public about how monopolies really worked, and why they were bad news for ordinary people. "'It is a practical demonstration of the present system of land-grabbing with all its usual outcomes and consequences,' she wrote in a political magazine. 'It might well have been called the "Game of Life," as it contains all the elements of success and failure in the real world, and the object is the same as the human race in general seem[s] to have, i.e., the accumulation of wealth.'"[484]

"Monopoly" is still a wildly popular board game, but, as it turns out, we didn't take the lessons from it that Ms. Magie might have wished we would. The parallels between the Gilded Age and our current Digital Age are remarkable—only this time the monopolies aren't in railroads

---

[484] Mary Pilon, " The Secret History of Monopoly: The Capitalist Board Game's Leftwing Origins," *The Guardian* (April 11, 2015), https://www.theguardian.com/lifeandstyle/2015/apr/11/secret-history-monopoly-capitalist-game-leftwing-origins.

and oil and steel, and it isn't only the United States that is home to these modern-day trust busters.

In the U.K., in April 2021, the "MailOnline/DailyMail.com [sued] Google for alleged anti-competitive behavior over the search giant's ability to exploit its dominance in the ad tech industry to harm rivals and manipulate ad auctions and news search results in a way that punishes online publishers."[485] The following June, the E.U. proposed a $425 million fine against Amazon for privacy violations,[486] and France fined Google £190 million for abusing its market dominance to sell online ads.[487] Even China, "the bulwark behind which American Big Tech has been hiding to avoid domestic regulation,"[488] is moving toward the "de-tycoonification"[489] of its tech sector.

Amid all of the international antitrust action, however, we can't go without mentioning that tech itself is also pursuing antitrust measures against tech. Epic Games, Inc., the creator of "Fortnite" and other popular games, has famously brought an antitrust case against Apple. Tim Sweeney, the co-founder and CEO of Epic, "has been a critic of Apple's business practices, saying the tech giant's fees result in higher costs to developers and unfairly restrict competition by not allowing competing

485 Emily Crane, "Daily Mail Files Anti-Trust Lawsuit Against Google," *Daily Mail* (April 20, 2021), https://www.dailymail.co.uk/news/article-9491607/Mail-files-antitrust-lawsuit-against-Google.html.

486 Sam Schechner, "Amazon Faces Possible $425 Million EU Privacy Fine," *The Wall Street Journal* (June 10, 2021), https://www.wsj.com/articles/amazon-faces-possible-425-million-eu-privacy-fine-11623332987?st=r4jgihjoxze43n-n&reflink=article_email_share.

487 Chris Pleasance, "France Fines Google," *Daily Mail* (June 7, 2021), https://www.dailymail.co.uk/news/article-9660017/France-fines-Google-190million-antitrust-case-online-ad-sales.html.

488 Tom Wheeler, "The Chinese government embraces tech industry competition," Brookings Institution (April 16, 2021), https://www.brookings.edu/blog/techtank/2021/04/16/the-chinese-government-embraces-tech-industry-competition/.

489 Ibid.

app stores on iOS devices."[490] Why is the battle joined between Epic and Apple riveting the attention of tech in the early months of 2021? Because the ultimate verdict could end up being, well, epic for the whole of tech: "Epic Games is asking the court to invalidate the entire business model behind the iOS ecosystem, seeking to bar Apple from requiring developers to use its in-app purchases for digital goods and services."[491]

To put a very fine point on it, the case between Epic and Apple illustrates in the most fundamental way how monopolies are bad for nearly everyone except the entity that actually owns the monopoly. In a nutshell, Apple's requirement that Epic's in-app purchases are made exclusively through Apple's store prevents Epic from selling through other vendors, and, in turn, this requirement deprives not only Epic but those other vendors—all smaller than Apple—from potential revenue that could come from customers whose digital lives take place outside of the Apple ecosystem. For Apple, and Apple only, this is a winning situation: if a customer wants to play a popular game like "Fortnight," they can do it only through Apple; and, not inconsequentially, revenues and growth for any competitor is depressed.

We don't want to single out Apple as having the only tech-on-tech problem. Other members of the Big Five have their own issues with monopolistic behavior. Take Amazon and the spat they had with smart-thermostat maker Ecobee in 2020. Amazon told Ecobee that it had to turn over data from its voice-enabled devices—which work with Alexa, Amazon's assistant—even when those devices weren't in use by customers. Ecobee declined, worrying they would be violating their customers' privacy—and they were worried the data would give Amazon insights the tech behemoth could put to use for competing products.

---

[490] "Epic vs. Apple Trial," *The Wall Street Journal* (May 2, 2021), https://www.wsj.com/articles/epic-vs-apple-trial-tim-cook-tim-sweeney-and-the-other-key-players-in-fortnite-makers-antitrust-lawsuit-11619969556?mod=djem10point.

[491] Ina Fried, "The trial that will decide the future of Apple's App Store," AXIOS (May 3, 2021), https://www.axios.com/apples-app-store-epic-games-fortnight-trial-3cb9eac4-83d6-4ad3-a0ad-c98bfcd65192.html?utm_source=newsletter&utm_medium=email&utm_campaign=newsletter_axioslogin&stream=top.

Amazon's response? That Ecobee's refusal could affect their ability to sell on Amazon.[492] *Nice little thermostat you've got there. Shame if something happened to it.* Negotiations to resolve the dispute are still ongoing as of this writing.

The cases I've outlined above don't, by any means, constitute an exhaustive list of antitrust actions being taken around the world by governments and private business—that's an exercise that would require a book unto itself, and one that required constant updating. Rather, the sampling of actions I've laid out demonstrate that the monopolization of tech presents a worldwide and increasingly urgent problem. Too much tech—too much money, too much control over the devices, software, and apps that we use daily to make our lives function more easily, too much of our private information now consolidated in the hands of too few. Paul Romer, once one of "Silicon Valley's favorite economists,"[493] concurs with this assessment, which "fits into a broader re-evaluation about the tech industry and government regulation among prominent economists. They see markets—search, social networks, online advertising, e-commerce— not behaving according to free-market theory. Monopoly or oligopoly seems to be the order of the day."[494]

So, just how much is in the hands of the new monopolists? Well, for example, since 2005, Facebook has acquired approximately ninety-two separate companies, including Instagram and WhatsApp, two of their former competitors. According to Tim Wu, professor of law, science, and technology at Columbia Law School, "It's against the law to buy your

---

[492] Dana Mattoli and Joe Flint, "How Amazon Strong-Arms Partners," *The Wall Street Journal* (April 14, 2021), https://www.wsj.com/articles/amazon-strong-arms-partners-across-multiple-businesses-11618410439?utm_source=The+Logic+Master+List&utm_campaign=07221ce6ff-Daily_Briefing_2021_Apr16_2_COPY_01&utm_medium=email&utm_term=0_325d5d3b52-07221ce6ff-275706585.

[493] Steve Lohr, "Once Tech's Favorite Economist, Now a Thorn in Its Side," *New York Times* (May 20, 2021), https://www.nytimes.com/2021/05/20/technology/tech-antitrust-paul-romer.html?referringSource=articleShare.

[494] Ibid.

competitors"; so, even barring new antitrust regulations, Facebook may already have broken standing antitrust law.[495]

Microsoft has acquired over 250 different companies since its initial IPO in 1986, including Skype, LinkedIn, and GitHub, and it holds stakes in nearly seventy companies, including Comcast, AT&T, WebMD, Best Buy—and Apple and Facebook. Google might hold the record for rapid acquisition: in one year, 2010–2011, it acquired, on average, more than one company per week. In 2015, Google restructured and created a new parent company for itself, Alphabet Inc, which owns over 250 companies, including the smartphone navigation platform Waze, YouTube, and FitBit. Amazon, for its part, owns approximately 108 companies, among them Zappos, Audible, Ring, which makes home security devices, and the grocery store chain Whole Foods.

What should the break-up of these companies—breaking their tentacle-like holds on competitors, as well as within industries as diverse as medicine and retail food—look like? Well, first, it should look like, and *be*, a carefully orchestrated set of events. In the case of Amazon, would we, as consumers, be better off with ten smaller Amazon-like companies that were keeping prices low because they were competing for our business? But would such a scenario work for a company such as Apple? "In any action, there's unintended consequences," says Wu. "If you imagine, for example, breaking up Apple into three little mini-Apples, there might be more competition, but it might be that people's phones don't work as well or, you know, they lose whatever magic mojo they had. So you don't want to actually make things worse for people, but that said, I think breakups have the positive effect of rebooting an industry, starting things afresh and frankly, historically, have often been better for people in the long run than anyone predicted."[496]

That said, regulation that requires a company to divest its interests in its competitors—think telling Facebook it can no longer own Instagram and WhatsApp—is an obvious first step, and requiring government

---

[495] "What Breaking Up Big Tech Might Look Like," All Tech Considered (October 21, 2019), https://prod-text.npr.org/772049931.

[496] Ibid.

scrutiny and approval before any new acquisition is made is the second. A third, related step is outlined in legislation introduced by Senators Amy Klobuchar (D-Minn.) and Charles E. Grassley (R-Iowa) in October, 2021, that would prohibit "so-called gatekeeper companies (or 'critical trading partners'), essential to other businesses' ability to reach their customers" from abusing their position.[497] In practical terms what this would mean is, for example, Amazon would not be allowed to promote its own product over a competitor's that is also being sold on Amazon *unless Amazon could prove that its own product held more value for the end consumer.* But is it possible to objectively prove that one box of tissues is better than another?

But just because these steps are obvious doesn't mean they're easy. Tech executives are among the top of the political donor class,[498] and they can have real stakes in agenda items such as the repeal of net neutrality and lobbying against antitrust legislation. Indeed, since six separate antitrust actions were introduced in the U.S. Congress as of June 2021, "Executives, lobbyists, and more than a dozen think tanks and advocacy groups paid by tech companies have swarmed Capitol offices, called and emailed lawmakers and their staff members, and written letters arguing there will be dire consequences for the industry and the country if the ideas become law."[499] It can't help the titans of tech to sleep well, either, that the Biden administration has named Lina Kahn as the new chair of

---

[497] "Finally, a promising piece of tech antitrust legislation in Congress," *New York Times*, Editorial Board, (October 24, 2021), https://www.washingtonpost.com/opinions/2021/10/24/finally-promising-piece-tech-antitrust-legislation-congress/.

[498] Ari Levy, "Here's the Final Tally of Where Tech Billionaires Donated for the 2020 Election," *CNBC* (November 2, 2020), https://www.cnbc.com/2020/11/02/tech-billionaire-2020-election-donations-final-tally.html.

[499] Cecilia Kang, David McCabe and Kenneth P. Vogel, "Tech Giants, Fearful of Proposals to Curb Them, Blitz Washington With Lobbying," *New York Times* (June 29, 2021), https://www.nytimes.com/2021/06/22/technology/amazon-apple-google-facebook-antitrust-bills.html?referringSource=articleShare.

the FTC: Ms. Kahn is a legal scholar whose work laid the foundation for the current antitrust efforts in Congress.[500]

During the last century's Gilded Age, the tycoons controlled the politicians, and it took true strength of will for Teddy Roosevelt to employ the Sherman Anti-Trust Act of 1890 to go up against the powerful railroad holding company, the Northern Securities Corporation, and bust that trust.[501] Considering this century's swirl of energy around new antitrust regulation, we're about to see if our contemporary politicians possess that same sort of strength.

○ ○ ○

In addition to regulating mergers and acquisitions through strengthened antitrust legislation, and stronger, proactive enforcement of those laws, there are other avenues of oversight that can and should be explored. On an international level, privacy issues fall under the United Nations Human Rights Office of the High Commissioner, Michelle Bachelet; and in 2015, the Human Rights Council appointed Professor Joseph Cannataci of Malta as the first Special Rapporteur on the right to privacy. "Article 12 of the Universal Declaration of Human Rights and Article 17 of the International Covenant on Civil and Political Rights provide that no one shall be subjected to arbitrary or unlawful interference with his or her privacy, family, home or correspondence, nor to unlawful attacks on his or her honor and reputation."[502] The most recent resolution on the right to privacy in the digital age was adopted in September of 2019, affirming that the same rights to which people are entitled offline must also be protected online.

---

[500] Ibid.

[501] "The Sherman Act," *Theodore Roosevelt Center, Dickinson State University*, https://www.theodorerooseveltcenter.org/Learn-About-TR/TR-Encyclopedia/Capitalism-and-Labor/The-Sherman-Act.aspx.

[502] "International Standards Relating to Digital Privacy," *United Nations Human Rights,* https://www.ohchr.org/EN/Issues/DigitalAge/Pages/International StandardsDigitalPrivacy.aspx.

This is an excellent start, but why stop there? Why not a UN High Commissioner on Tech Influence? Human lives all over the world are undergoing datification at the hands of corporations that are not yet too big to fail, but might soon be too big to stop. If the United Nations really is "the central institution in the world for promoting international peace and security and the rule of law"[503]—as indeed it is—then the full force of its authority needs to be directed toward addressing what is, by any measure, a global crisis. Like the United Nations High Commissioner for Human Rights, and the High Commissioner for Refugees, the one I propose will deal exclusively with a global problem that urgently requires focused and dedicated attention.

Further, if, as the magazine *Wired* put it, "The world is no longer dominated by nation-states alone. We have moved into the non-state, net-state era,"[504] then the logical next step is to call for the appointment of tech ambassadors, on the order of what Denmark did by appointing Casper Klynge the world's very first tech ambassador—the same Casper Klynge who now works, I remind you, as Microsoft's UN liaison. Right now, the biggest of the Big Tech companies are operating in the global landscape with the advantage of annual budgets and more reach than some actual geographical countries. "A non-nation-state, Facebook, just topped 2 billion users—more than a quarter of the world's population, surpassing even China's population by almost 40 percent."[505] It is time nations begin talking tech together on a diplomatic level.

Diplomacy, the skill of managing international relations, revolves around conducting negotiations between or among countries and, often, corporations from those countries, in order to secure long-term security

---

[503] "New York City and the United Nations: Toward a Renewed Relationship," *New York City Bar* (July 1, 2002), https://www.nycbar.org/member-and-career-services/committees/reports-listing/reports/detail/new-york-city-and-the-united-nations-toward-a-renewed-relationship.

[504] Alexis Wichowski, " Net States Rule the World. We Need to Recognize Their Power," *Wired* (November 4, 2017), https://www.wired.com/story/net-states-rule-the-world-we-need-to-recognize-their-power/.

[505] Ibid.

and cooperation goals. As current events evidence, these net-state actors have been and continue to be powerful forces on the worldwide stage, and we should begin treating them with the diplomatic gravity they warrant—for the good of world peace and the good of the world's economies, as well as for the good of the net states themselves. The Editorial Board of the *New York Times* opined in September 2021 that "Apple and Google are showing Putin just how much he can get away with"[506] prior to that month's Russian elections. Specifically, they dropped a pro-Navalny app—named for opposition figure Alexei Navalny—after Putin's regime moved from issuing verbal threats to arresting employees of the tech firms to sending armed men to Google's Moscow offices. And, in Afghanistan, resistance leaders report that the Taliban, since retaking control of the country, are forcing internet outages, resulting in complete communications blackouts.[507] These incidents demonstrate how powerful these tech giants are—clearly both Putin and the Taliban place a high value on the ability of high-tech communications to disrupt their control in their respective countries. They also demonstrate, however, the vulnerability of the citizens of those countries to loss of communication technologies. And the vulnerability of the employees of the tech companies to armed and violent threat. And the vulnerability of the world to regimes that are determined to "isolate protesters and quash dissent."[508] These are just two examples of the urgency for diplomatic intervention vis-à-vis Big Tech.

The United Nations Security Council "has established a wide-variety [sic] of Commissions to handle a variety of tasks related to the maintenance of international peace and security. Commissions have been created with different structures and a wide variety of mandates including

---

[506] "Apple and Google are showing Putin just how much he can get away with," Editorial Board, *Washington Post* (September 18, 2021), https://www.washingtonpost.com/opinions/2021/09/18/apple-google-are-showing-putin-just-how-much-he-can-get-away-with/.

[507] Shannon Vavra and Diana Falzone, "This is Why the Taliban Keeps F*cking Up the Internet," *The Daily Beast* (September 16, 2021), https://www.thedailybeast.com/this-is-why-the-taliban-keeps-fcking-up-afghanistans-internet?ref=home.

[508] Ibid.

investigation, mediation, or administering compensation."[509] Given the myriad ways in which tech has already transformed our world, and the enormity of the ways in which it is poised to transform our world in the coming years—from impacting national economies and thus the stability of every nation on the globe, to its part in the way international conflicts will be handled and cyberwarfare will be waged in the future—establishing a UN Commission on Technology is not only prudent, but likely paramount to global peace.

Further, nations should follow Denmark's lead and name tech ambassadors to function as part of the diplomatic corps accredited to each particular country, or body. Diplomats—an overarching word to describe anyone who works in the foreign service—act as representatives of their home countries in relation to the country in which they are posted, facilitating relations with that country politically, culturally, economically, and militarily. This is especially the case concerning negotiations of treaties and pacts, and particularly in times of crisis. As tech plays an ever-expanding role in each of these areas of diplomatic concern, it only makes sense that countries have designated tech ambassadors—experts in technology with their boots on the ground, so to speak, who can understand, explain, intervene, and negotiate with our friends as well as our adversaries in matters that will increasingly relate to or contain a component dependent on an aspect of technology.

On a national level within the United States, I propose the creation of a Technology Fairness Commission to oversee matters from antitrust to privacy regulations involving Big Tech. My primary purpose for this proposal is speed. The antitrust cases currently making their way through United States courts will, as these sort of complicated cases against well-financed defendants often do, take years to reach a verdict and to result in legislation or any other concrete actions, such as the break-ups of existing monopolies. In the meantime, Big Tech will keep acquiring new subsidiaries—including from among their competitors—and they'll keep making millions from the use of your private data.

---

[509] United Nations Security Council website, https://www.un.org/securitycouncil/content/repertoire/commissions-and-investigative-bodies.

A commission designed specifically to enforce existing laws and regulations could move immediately to abate some of Big Tech's most egregious behaviors. Agents from the newly minted Technology Fairness Commission could be embedded in the offices of Big Tech, so they become an on-site asset for Big Tech companies to, for example, review and advise concerning new acquisitions *before* the company violates a law. Within the commission, there should also be a False Claims Bureau dedicated to investigating suspicious online news feeds around issues of factual accuracy as well as point of origin. Dubious news items traced to the activity of foreign adversaries can be taken offline before they have a chance to spread and amplify, impacting U.S. policy or elections. Ideas can be planted like seeds on social media, watered by the reaction they get from users, transplanted with each new "share," then begin to invade offline lives, until you have a veritable forest of falsehood, an invasion of undergrowth, an insurrection. As Joan Donovan has said, "Responsiveness is the core feature of social media," and our ability to respond to cyberattacks by foreign adversaries—whether they come from ransomware hackers or social media trolls—needs to be enhanced, quickly and exponentially.

There is plenty of precedent for the creation of such a proposed agency. In the financial world, the Federal Deposit Insurance Corporation (FDIC) provides deposit insurance, guaranteeing deposits to $250,000 for member banks, analyzes the stability of financial institutions, and manages failed banks. The Consumer Financial Protection Bureau (CFPB) oversees finance-related products and services aimed at consumers. The Securities and Exchange Commission (SEC), among the most powerful of regulatory agencies, enforces federal securities laws and regulates much of the securities industry, including U.S. stock exchanges. The Occupational Safety and Health Administration (OSHA) develops and enforces federal standards and regulations ensuring worker safety. The Food and Drug Administration (FDA) administers federal food safety laws, drug safety, and cosmetics safety. The Federal Communications Commission (FCC) regulates interstate and foreign communication by way of radio, telephone, telegraph, and television—but not, you will note, by way of computer or other digital devices. The Federal Trade Commission (FTC) is charged with ensuring free and fair market competition and protects consumers from deceptive practices. I could go on, but you get the point: I'm proposing a traditional solution to regulate a new industry that, to date, has been operating as if they're living in the Wild West and there's no sheriff in town.

And, as we're in the realm of what nation-states can do to check the power of net-states, one more thing: we can switch from using Google and Bing and the like as our default search engines and start using services such as DuckDuckGo.[510] DuckDuckGo works just like any other browser to which you've become accustomed, but it doesn't store your personal information. It still generates revenue through selling ads but, as DuckDuckGo explains it: "On other search engines, ads are based on profiles compiled about you using your personal information like search, browsing, and purchase history. Since we don't collect that information, search ads on DuckDuckGo are based on the search results page you are viewing, not on you as a person. For example, if you search for cars, we'll show you ads about cars."[511]

But let's also imagine the possibility of the creation of a public, fee-free, and ad-free search engine—the NPR of search; call it "Oogle"—that's available as an alternative to existing, for-profit services. Is it wholly realistic to imagine a truly neutral algorithm that bases the results it serves to users on "just the facts, ma'am"? It could be, if we remove the user's personal data from that algorithm's parameters. If I search for the "lunch menus" of my local restaurants, the one place near me that serves cashew chicken salad is likely going to appear at the top of my results; if the algorithm wasn't aware of my affinity for this dish, I could very well end up making a more adventurous lunch choice from time to time. Now, I realize this is an innocuous example, but think about the wealth of non-biased information a user could receive if a search algorithm didn't already know the deeply personal details it now uses to keep her or him

---

[510] DuckDuckGo website, https://duckduckgo.com/?va=b&t=hr.
[511] Ibid.

firmly within an age-related, or socioeconomic, or politically-infused filter bubble.

○ ○ ○

Moving from what we should expect governments around the world to do to check tech, we come to the actions *we*, ourselves, can take to limit the invasiveness—and, perhaps, at least some of the pervasiveness—of tech in our lives, to make our user experiences pleasant, efficient, and personalized, but on *our terms*, not Big Tech's.

We start with the Internet Users' Bill of Rights, an idea first proposed by Sir Tim Berners-Lee. He is a British computer scientist who, in 1989, drafted a proposal for the entity that has since become the World Wide Web, and so is generally given credit as the WWW's inventor. In 2014, on the twenty-fifth anniversary of that initial proposal, Berners-Lee drafted another one, an Internet User's Bill of Rights—an "Online Magna Carta" that would assure neither governments nor corporations would be able to control or curb its neutrality in the future. "I believe we can build a Web that is truly for everyone: one that is accessible to all, from any device, and one that empowers all of us to achieve our dignity, rights and potential as humans."[512] The six key, basic principles Berners-Lee outlined in his "Magna Carta" are:

1. Accessibility
2. Affordability
3. Privacy
4. Freedom of expression
5. Diverse, decentralized and open platform
6. Net neutrality for users and content alike

---

[512] Paul Vale, "Tim Berners-Lee Calls for 'Online Magna Carta' To Protect Net Neutrality On 25th Anniversary," *Huffington Post* (November 3, 2014), https://www.huffingtonpost.co.uk/2014/03/11/tim-berners-lee-online-magna-carta_n_4945595.html.

You'll note that these are principles for which I've advocated within these pages, and you'll also note that most of them involve the cooperation and intervention of both of the very entities—governments and corporations—that Berners-Lee believes should not be able to place restraints on our free use of the internet. We cannot achieve, for example, affordability without placing restrictions on what broadband providers can charge as a fair price for their service; we cannot achieve net neutrality without governments that will stand in the way of Big Tech companies receiving preferential treatment over their smaller competitors.

This said, not all solutions rest at the macro level. There are many ways in which we can intervene in the micro on our own behalf to enhance our user experience and protect our privacy.

First among these options is to opt out of cookies, and thus the targeted ads they send our way. This option sounds simpler and more straightforward than it actually is, though. The foremost reason is because, as we've learned, when you do a great deal of opting out—at least in the way the internet is currently configured to work, with lots of third-party and zombie cookies—you lose a certain amount of functionality, and your user experience can become less pleasant. The second reason belies the first, because there is simply no one way to permanently opt out of being a target.

The first stop on your quest to opt-out might be AdChoices, a program organized by a group of online advertisers called the Digital Advertising Alliance. This group is behind a whole host of the ads that target us, and AdChoices is their attempt to regulate themselves. This attempt is certainly admirable, but there's a catch: the advertiser that is sponsoring any given ad has to agree to be a part of the opt-out program, thus allowing you the choice to view their ad or not.

Now, there are a few reasons an advertiser might go along and give such permission—chief among them being that if you opt out of their ad, you weren't likely to be a lucrative customer anyway, so why bother spending advertising dollars to try to reach you again and again when, by letting you off the hook, they can spend those same dollars on a more "gettable" customer? That said, if you see a blue triangle at the top of an ad, that means the sponsor of the ad has opted-in to allow you to opt-out;

by clicking on the triangle, you can get more information about why you were targeted, and you can change settings so you will no longer receive ads from that company.

Within AdChoices there is also a tool called WebChoices, which gives you a little more latitude, meaning that instead of opting out of one ad at a time, you can opt out of receiving ads from participating companies en masse. One of the downsides with this tool, however, is that you have to opt-out for each browser you use, not only your primary one. That means you'll need to repeat the process for each device you use—your computer, your laptop, your phone, and your tablet, and again for each browser you use on each device. Currently, only about two hundred companies have opted in to allow you to opt-out—and exercising your opt-out doesn't mean you won't see *any* ads, only that those ads won't be personalized, based on your personal data. But among the companies are some big names from the United States, Canada, and the E.U.: AT&T, Bloomberg, Comcast, General Motors, Procter & Gamble, Conde Nast, Dow Jones, as well as Facebook and Microsoft.

You can also selectively opt-out on a per-platform basis. On Facebook, for example, under "Account Settings" there is a tab for "Ads." In this tab you'll find a trove of your personal information that advertisers can see and use to target their ads. Though you can't entirely opt out, you can customize your ad experience by clicking on each category. You can also go to the "Interests" tab and select content for which you no longer wish to receive advertisements. This means that if last year you were, say, researching the best snow blower to get your father-in-law for his birthday, you can indicate to Facebook that you've made that particular purchase and are no longer interested in seeing ads for that particular item, or from any one particular company.

Instagram ads, Google ads, Twitter ads—all have the same sort of feature through which you can opt out of receiving specific ads, or specific categories of ads. Navigating to any of these platform-specific opt-out options is certainly not intuitive, and Instagram's opt-out feature is the most cumbersome, in that you cannot opt out of categories of ads but *must* opt out of each actual ad individually. Keep in mind that, even when you have successfully completed all the steps to opt-out of advertisements in these ways, you can't count on your opt-out being permanent. As soon as you visit a site that uses cookies, you've made a new pathway for the same old advertisers to show up again in your newsfeed. If your father-in-law's snow blower breaks and you need to buy him a new one, you are once again opening the door and inviting the snow blower manufacturers to come knocking.

If all of this seems complicated and cumbersome, you're right to feel that way—because it is, and that's why most people don't bother opting out. A study conducted by *Consumer Reports* found that at least one in three test subjects failed to find the "Do Not Sell," or DNS link that would prevent a site from selling their private information.[513] Another study, this one conducted by DataGrail in the first two quarters of 2020, found only 84 DNS requests per million user requests.[514] In response to this difficulty—and to a provision of the California Consumer Privacy Act that specifies businesses must respect a global or universal opt-out sent by a device or a browser—a group of "privacy-focused tech companies, nonprofits, and publishers, including the *New York Times*, the Electronic Frontier Foundation, and the search engine and browser DuckDuckGo"[515] is beta testing a universal privacy control that would allow the consumer to flip one simple opt-out control on his or her mobile phone, or within his or her browser, to effectively, efficiently, and completely opt-out. Called OptMeowt, this privacy feature is currently available as a free browser enhancer for Chrome and Firefox. Its source code is open to the public on GitHub, so users can send feedback to developers. "For California residents, the global privacy control, if enforced by the attorney general, would have a very different effect than existing privacy controls such as third-party cookie blockers. Those settings have

---

[513] Maureen Mahoney, "California Consumer Privacy Act," https://advocacy.consumerreports.org/wp-content/uploads/2020/09/CR_CCPA-Are-Consumers-Digital-Rights-Protected_092020_vf.pdf.

[514] DataGrail website, https://www.datagrail.io/CCPA_DSAR_H12020/.

[515] Gilad Edelman, "'Do Not Track' Is Back, and This Time It Might Work," *Wired* (October 7, 2020), https://www.wired.com/story/global-privacy-control-launches-do-not-track-is-back/.

no power over what a website or app does with the data it collects directly from you. The global control, by contrast, would issue a legally binding order that, if violated, would be punishable by major fines."[516]

Cookies were created in the first place to make our user experience on the internet easier and more seamless. Whatever nefarious mutations they've undergone in the interim between their invention and now, they do have their purposes—such as keeping track of items in your shopping cart as you move from page to page on a retailer's website. Disabling cookies can change and even frustrate or impair your user experience by no longer performing tasks they were designed to perform, and to which you have grown accustomed to having done for you as you shop or work online.

o o o

Speaking of opting out, if you own an Amazon device, such as an Echo Dot that you purchased after 2018, you might want to switch off your Sidewalk. "Sidewalk" is an extensive new wireless network that is entirely controlled by Amazon. The capability to activate this network has been built into Amazon devices since 2018, but they lay dormant—until June of 2021, when Amazon decided to turn them on.[517]

The stated intent of Sidewalk is to make our wireless devices work more efficiently. For example, most home Wi-Fi systems work only within the home itself and do not extend to outdoor areas, like backyards and, well, *sidewalks*. With an Amazon Echo, however, your existing private internet connection is transformed into what the company calls a "Sidewalk Bridge" that uses Bluetooth to connect nearby devices, and a different, separate signal to connect to devices up to half a mile away. In doing so, "Sidewalk authorizes your Echo to share a portion of your home's internet

---

[516] Ibid.
[517] Sara Morrison, "Where the (Amazon) Sidewalk Ends," *Vox* (June 8, 2021), https://www.vox.com/recode/22516285/amazon-sidewalk-echo-tile-ring.

bandwidth. It's up to 500 megabytes per month—the rough equivalent of more than 150 cellphone photos."[518]

The concerns associated with this new Sidewalk network are many. How will Amazon use this network to more aggressively surveil its customers and collect even more private information about us? What sort of additional security risks is Amazon creating for us by opening up this new portal to our homes? Can this new network also be used by Amazon to, for instance, track its package delivery, and if so, shouldn't Amazon be paying for it? "Which raises the question, then, if Amazon is using Sidewalk to collect even more of our personal data *and* track its own package deliveries: Shouldn't Amazon be paying *us*?"[519] (The emphasis is mine but the short answer to the question is, "Yes.")

Meantime, you might want to consider going into the settings on your Echo and turning off your Sidewalk.

o o o

Another avenue to consider as you're working to better secure your private information are VPNs. A virtual private network, or VPN, is a sort of software program that establishes a secure connection, called a "VPN tunnel," between your computer and a remote VPN server. In essence, this tunnel creates a private network within the connection that already exists within your public server. Among its features, the connection to the VPN encrypts your information, hiding your IP address so your identity is hidden from any website you might visit, and making it safe for you to use public WiFi. It also hides your web activity from your primary Internet Service Provider (ISP).

These are all among the benefits of installing a VPN, though keep in mind: they do not make you completely *anonymous* on the internet. For instance, a VPN will obscure your actual geographical location; but a

---

[518] Geoffrey A. Fowler, "Amazon may be sharing your Internet connection with the neighbors. Here's how to turn it off.", *Washington Post* (June 8, 2021), https://www.washingtonpost.com/technology/2021/06/07/amazon-sidewalk-network/#HOWTO.

[519] Ibid.

sophisticated hacker—or, say, the FBI—will always be able to track web activity back to *you*, if not your exact location.

Among the disadvantages of VPNs are that they do increase your data usage, and they generally decrease the speed of your internet connections. Further, not all VPN operations are scrupulous, and a dodgy one can leave you vulnerable to malware and hackers. So, doing your due diligence before settling on a VPN to install is critical. Here's a tip: VPNs are a paid service, typically costing the user between $2 and $15 per month. As you do your research, you might find there are quite a lot of "free" VPN services on offer. Be wary of such "free" services, as these are often hackers looking for an easy way into your computer—and your data.

o o o

Among the most pressing concerns that many of us feel around the dangers of tech is the responsibility to protect our children from the online abuse and predation that have, over time, grown ever more prevalent.

In Upton, California, a man was arrested after a ten-month investigation for having images of child pornography stored on his mobile phone—inside an app made to look like a standard calculator.[520]

A July 2021 study by the Mozilla Foundation, the nonprofit behind the Firefox browser, found that 71 percent of videos flagged by a team of 37,000 YouTube users "included COVID-19 misinformation, political conspiracy theories, and both violent and graphic content, including

---

[520] Norman Miller, "Federal Agents Say Upton Man Had Child Pornographic Images in His Phone," *MetroWest Daily News* (July 15, 2021), https://www. metrowestdailynews.com/story/news/2021/07/15/upton-man-faces-20-years-federal-prison-child-porn-possession-charge/7977148002/.

sexual content that appeared to be cartoons for children"[521]—such as a sexualized parody of childhood favorite, *Toy Story*.[522]

A Texas teen shot herself after months of receiving abusive texts by way of an untraceable smartphone app[523]—just one of the hundreds of sad stories that connect cyberbullying with teen suicide, suicide attempts, and suicidal ideation.

Safe Horizon, an organization dedicated to fighting human trafficking, estimates that almost twenty-one million people worldwide are trafficked.[524] Save the Children reports that over a quarter of all trafficking victims are children, and that two out of every three children is a girl. These children are abducted for the purposes of forced labor or sexual exploitation.[525]

When we reflect on the many ways in which children can be harmed in an online environment, that responsibility can feel especially overwhelming. As with any problem, however, the solution is sunlight; indeed, the U.S. Department of Homeland Security states it most succinctly: "Predators and human traffickers can gain access to victims online because people are not always aware of how dangerous online

---

[521] Brad Zadrozny, " YouTube's Recommendations Still Push Harmful Videos, Crowdsourced Study Finds," *NBC* (July 7, 2021), https://www.nbcnews.com/tech/tech-news/youtubes-recommendations-still-push-harmful-videos-crowdsourced-study-rcna1355.

[522] Tripp Mickle, "YouTube's Search Algorithm Directs Viewers to False and Sexualized Videos, Study Finds," *The Wall Street Journal* (July 7, 2021), https://www.wsj.com/articles/youtubes-search-algorithm-directs-viewers-to-false-and-sexualized-videos-study-finds-11625644803?st=y7hxxa6x7faonqk&reflink=article_email_share.

[523] "Cyberbullying Pushed Texas Teen to Commit Suicide, Family Says," *CBS* (December 2, 2016), https://www.cbsnews.com/news/cyberbullying-pushed-texas-teen-commit-suicide-family/.

[524] Safe Horizon website, https://www.safehorizon.org/get-informed/human-trafficking-statistics-facts/#our-impact/.

[525] "The Fight Against Child Trafficking," *Save the Children*, https://www.savethechildren.org/us/charity-stories/child-trafficking-awareness.

environments can be or how to keep themselves safe,"[526] and it offers these online safety tips to kids and the adults who love them:

- Never share pictures of yourself online that you wouldn't want to be seen by your family, teachers, or a total stranger.
- Set user profiles to private, so only real friends can get access. Know who you're chatting with—a "friend" is not always a friend.
- Treat people online as you would in person: be polite!
- Don't share personal information online, such as your full name, school, address or phone number, or user passwords.
- Don't meet up in person with anyone you met online.
- Report suspected abuse to law enforcement or a trusted adult.

The government of Vietnam has attempted a more proactive approach, instituting national guidelines for social media behavior: "Social media users are encouraged to promote the beauty of Vietnam's scenery, people and culture, and spread good stories about good people," reads the code.[527] The idea is a lovely one, and I'm not saying that similar guidelines in other countries would have *no* effect. But, as Reuters reported, it was not clear to what extent the Vietnamese government's decision was legally binding, and the decision was not accompanied by any indication of how it would be enforced.

Until we can ask people to treat each other with kindness and honesty, and come away with the sincere belief that they will indeed act with decency, there are a number of parental control apps[528] that can help parents

---

[526] Department of Homeland Security website, https://www.dhs.gov/blue-campaign/online-safety.

[527] "Vietnam Introduces Nationwide Code of Conduct for Social Media," *Reuters* (June 18, 2021), https://news.trust.org/item/20210618040727-kltxd.

[528] "10 Best Parental Control Apps of 2021," *Consumers Advocate* (November 2, 2021), https://www.consumersadvocate.org/parental-control-apps/a/best-parental-control-apps?pd=true&keyword=%2Bmonitor%20%2Bkids%20%2Bonline&bca_campaignid=333739434&bca_adgroupid=12149606544 26903&bca_matchtype=p&bca_network=o&bca_device=c&msclkid=bdf8b-20713d319a0110cd2f76d1b1eea.

to protect their kids. Rated at the top of that list is an app called "Bark."[529] Among the services provided by this app are: social media monitoring, which tracks conversations and content on the platforms your kids use most, like Snapchat, Instagram, YouTube, Twitter, Facebook, and more; monitoring for your kids' emails and texts, including screening any photos and videos sent or received for inappropriate content that might indicate interaction with a predator, receipt of adult content, sexting, drug use, suicidal thoughts or ideation, and cyberbullying; a screen-time manager, limiting when your kids can access the web on their devices, and web filtering, to prevent them from visiting inappropriate websites. Bark is on duty 24/7. It sends automatic alerts to parents when it detects potential danger—and its alerts include recommendations from child psychologists and other experts for dealing with them.

If all this sounds a bit intrusive, well, it might be; but it is not out of proportion with the scope of the danger the internet presents to our kids every day. Think of it this way: you can monitor your child's online activity yourself, which means essentially following her around and leaning over her shoulder whenever she logs on to a device—not to mention keeping up with teen culture, so you know what you're looking for as you lean over her shoulder to scan her emails or the websites she's visiting. Or you can use a service like Bark that will do the monitoring for you, unobtrusively, protecting your kid's privacy, even from you, and alerting you only when it senses an area for potential concern.

o o o

The approaches I've described so far revolve around twenty-first-century housekeeping tasks that are necessary to keep your family safe and your digital life tidy, or around larger institutions taking action to regulate tech. These last few suggestions involve something more personal: a baseline reassessment of your relationship with tech, so it can become a healthier and more sustainable one.

---

[529] Bark website, https://www.bark.us.

The first question to ask yourself as you do this reevaluation is a blunt one: Is it time to break up with Mark Zuckerberg? Or, at least, is it time to take a little break?

A friend of mine has a long history of upholding the tradition of "Sunday dinner" for family and friends. The faces around the table each week change frequently, depending on who is out of town on a business trip or who is home visiting from college. But each guest knows the host's one truly hard and fast rule: "No electronics at the table!" They respect it—and even the teenagers seem to really enjoy the scant hour or two of convivial conversation as a respite from their phones, and from the distractions and worries that come with constant checks of email and social media. What would happen if we would all take a regular break from our own electronics? If we instituted a movement on the order of "Facebook Free Fridays," during which the whole world simply did not check its Facebook accounts for one, brief, twenty-four hour-period, once a week? What would that look like?

We have some baseline for understanding such a scenario thanks to the six-hour outage experienced by Facebook and its apps, including WhatsApp and Instagram, on October 4, 2021, but let's break it down anyway. Initially, it might look like millions of people missing the dopamine rush that comes with accumulating "likes" and comments on a post about the corruption that's rampant on the *other* side of the political spectrum, or about the cuteness of their cat. But once that first shock has worn off, what would we all do with the time and energy we weren't pouring out on Facebook for the day? What would you do with those extra assets once you freed them from social media? Work on a hobby? Read a book? Take a walk or head to the gym? Cook a Sunday dinner? What do you imagine your emotional or your mental state might be if you removed the distractions and the worries of social media for just one day? Would you feel calmer? More productive? More connected to the actual people in your lives or around your dinner table?

And, no less importantly, what would a Facebook Free Friday look like to Mark Zuckerberg? At the end of the day, as ethicist Tristan Harris asks us to remember, Facebook's stock price is dependent on keeping people's attention.[530] What would the whole world forgetting about Facebook and withholding that attention for just one day each week do to Facebook's revenue stream? Would it capture the notice of the folks who work at Facebook if it wasn't the government but us—Facebook's *users*— who were upending their business model? What would it say about social media's clout if we purposefully and mindfully limited the power that tech has over us? In October 2021, Mark Zuckerberg took a $6 billion loss thanks to one, short, six-hour outage.[531] Now, to take the broad view, the outage had serious economic impact beyond Mr. Zuckerberg; think, for example, of the small businesses who use Facebook, and Facebook Marketplace, to sell goods and services, and the income they lost because the social media went down. "As the *New York Times* reported, some people were literally unable to communicate with health care workers and some businesses lost total contact with customers because Facebook products have so completely monopolized communications in their communities. [Senator Elizabeth] Warren accused Big Tech of using 'their resources and control over the way we use the Internet to squash small businesses and innovation,' but even she probably didn't imagine that the problem is so serious that an empanada seller in Colombia would lose a day of sales without WhatsApp."[532]

Think also, however, of how Mr. Zuckerberg would react if we users purposefully and consistently denied him $6 billion each and every week. Do you think, in this case, we would get his attention and he might really come around to reshaping his platform in *our* image, as one that is more humane and civic-minded?

---

[530] Tristan Harris, "An Arms Race for Your Attention," eCorner, Stanford University (November 5, 2017), https://ecorner.stanford.edu/clips/an-arms-race-for-your-attention/.

[531] Scott Carpenter, "Zuckerberg Loses $6 Billion in Hours as Facebook Plunges," *Bloomberg* (October 4, 2021), https://www.bloomberg.com/news/articles/2021-10-04/zuckerberg-loses-7-billion-in-hours-as-facebook-plunges.

[532] Amanda Marcotte, "Facebook Outage Proves Elizabeth Warren Is Right: It's Time to Break Up Big Tech," *Salon* (October 2, 2021), https://www.salon.com/2021/10/05/facebooks-outage-proves-elizabeth-warren-right-its-time-to-break-up-big-tech/.

The second part of reassessing our relationship with social media is to do a self-examination of our filter bubbles—and then to pop them. As we've discussed at length within these pages, social media is designed to keep your attention through the use of algorithms that coordinate what they know about your personal preferences to present you with a customized slice of reality. What would happen if you threw the algorithm a curve ball?

What if we proactively decided to diversify the input we receive? How would that affect our understanding of the world, our insights into problem-solving, and our empathy for our fellow human beings? And, not peripherally, how would Facebook itself react if, suddenly, they found that their algorithms weren't working as efficiently to extract from our personal information our needs and desires and lay them at the feet of social media advertisers? How would the advertisers react if, suddenly, we decided to present ourselves again as whole, complicated human beings instead of a bunch of easily targeted data points?

But how do you break a filter bubble and integrate new and even novel worldviews? Here's a short, sweet starter list.

1. Read a newspaper or magazine, watch a television show, or listen to a podcast that doesn't align with your present-day perspective. I'm not suggesting that if you're a die-hard progressive who listens regularly to *Pod Save America* that you remove that podcast from your playlist and jump right into an exclusive and steady diet of *The Ben Shapiro Show*. I'm suggesting that you supplement your news diet with reliable sources from the side of the argument to which you've rarely, maybe even never before, exposed yourself.
2. Join a Facebook group comprised of people who have a different take than you do about a subject in which you're interested, or follow a public figure who is impacting current events, and be respectful as you learn from each other.
3. Follow someone on Twitter who will provide a better understanding of what is motivating the folks who disagree with your point of view.

The caveat here is that doing these simple, obvious things is *hard*. We live in a sharply divided world, and people who differ in opinion, but are also respectful of others when they differ, can seem as if they're in short supply. This wasn't always the case, of course. The United States has enjoyed many sustained periods of broad national unity, generally in times of crisis during which a sense of common purpose was necessary to overcoming adversity, and the sentiment of concord was a comfort to ragged spirits.

Franklin Delano Roosevelt—known to history as FDR, as one of our greatest presidents—presided over two of the most searing crises in American history: the Great Depression and World War II. During his time in office, from 1933 until 1944, FDR conducted a regular series of "Fireside Chats" that were broadcast in real time over the nation's radio networks, with FDR seated not before a fireplace but in the Oval Office, behind a desk crowded with microphones. He used the chats to put his proposals for economic recovery directly to the American people, and to counter arguments against the New Deal. To inform the people about the status of the war, and to address rumors about its prosecution. To reassure the public that he was an attentive, engaged, thoughtful, and empathetic leader through a time of despair and uncertainty.

These chats were appointment listening for millions of Americans. Families gathered around their radios to hear what their president had to say. Folks who didn't own radios could gather at movie theatres and other public venues, in towns small and large across the nation, where the theatre owners would preempt any scheduled picture show in order to broadcast a presidential chat. A person walking down any city street during one of FDR's chats would hear his message coming across the airwaves from the radio sets of people who lived all along the street. Through these chats, FDR galvanized the public behind his ideas, which both enabled him to make progress in implementing them and offered a reason for optimism to a country sorely in need of one. The chats also did something else: they changed the relationship between a president and the public. The president and an overwhelming majority of that public were on the same page, unified in their goals—Get people back to work!

Defeat fascism!—and they were doing it with, if not cheerful, then solid and sustained confidence.

We no longer live in that world.

The trick, then, is to consider our sources very carefully and be judicious about which ones we choose to take to heart. Remember that in our digital age, anyone with an internet connection can have a public opinion—and a Twitter following in the millions—and merely possessing either or both of those things doesn't make that person credible. Evaluate your potential sources through a healthily skeptical lens. Who is the author, and what are her or his credentials? What sources does the author use in evaluating her own data, and do *those* sources stand up to scrutiny? Are they specific—does the author cite a particular study or studies to support his takeaway, or are the citations more ambiguous, more on the order of "many people say" or "recent events prove"? News happens fast these days—is your chosen source up-to-date? Who has provided reviews of the author's work, and are those reviewers themselves reliable?

I'm obliged to point out that I'm not saying here that we should be fact-checking only news sources and politicians, but the actual reviews that more often than not accompany products for sale and services for hire on online. Reviews for products and for services—books, lawn mowers, humidifiers, hairdressing salons—are important to a seller because they burnish the credibility of those things on offer. Reviews from buyers who weigh in on the quality of wares, or the courtesy of the customer service, provide the social proof that allows potential buyers to have a level of comfort in those things on which they are spending their own money.

Robert Cialdini—the Regents' Professor Emeritus of Psychology and Marketing at Arizona State University, and author of the seminal 1984 book, *Influence: The Psychology of Persuasion*—says: "Social proof works by reducing uncertainty. We live in what is the most information-overloaded, stimulus-saturated environment that's ever existed on our planet. We have all these choices, all these opportunities to move in one direction or another, all this information, and it leads us to be uncertain about the best choice to make. One way we can reduce that uncertainty is to ask ourselves, 'What are the people around me, like me, doing in this situation?' It's something that we call *peer-suasion*. Instead of persuasion, *peer*-suasion, which turns out to be more effective than simple persuasion. If you can get people to see that a lot of others are experiencing your product or service or idea in a positive way, well, you've reduced their uncertainty about the likelihood it's going to be positive for them too."[533] Indeed, in a 2021 survey, a full "78% of people said that product reviews played a big role in their purchase decisions."[534] And those star, rave ratings? The American Economic Review has found that, "The aggregate effect of star ratings on consumer surplus is, in our baseline estimates, more than ten times the effect of traditional review outlets."[535]

But just how reliable are those glowing reviews? How can you tell if they're *real*?

"Bogus product reviews are an epidemic, according to Saoud Khalifah, founder and CEO of Fakespot,[536] a site that ferrets out fake reviews. Khalifah says many product reviews are not real—and he has the data to prove it. 'Companies constantly plant positive reviews of their own products and sully competitors' products with negative reviews,' he says. As a result, many of the ratings you read online aren't credible. For example, up to 70 percent of the reviews on Amazon are not real, he says."[537] The problem is so rampant that it's attracting regulatory scrutiny. In the U.K., the Competition and Markets Authority (CMA) is investigating concerns that "Amazon and Alphabet's Google haven't been doing

---

[533] Robert Cialdini, "Influence: The Psychology of Persuasion," *YouTube* (May 24, 2021), https://www.youtube.com/watch?v=5Fv7s5sVVxA.

[534] Jason Cohen, "How to Spot a Fake Review on Amazon," *PC Magazine* (June 21, 2021), https://www.pcmag.com/how-to/spot-a-fake-review-on-amazon.

[535] Imke Reimers and Joel Waldfogel, "Digitization and Pre-Purchase Information," *American Economic Association* (June 6, 2021), https://www.aeaweb.org/articles?campaign_id=4&emc=edit_dk_20210618&id=10.1257%-2Faer.20200153&instance_id=33313&nl=dealbook°i_id=59648700&segment_id=61049&te=1&user_id=fa06607e4a3b78b585d4ddf2707df285.

[536] Fakespot website, https://www.fakespot.com.

[537] Christopher Elliott, "This Is Why You Should Not Trust Online Reviews," *Forbes* (November 21, 2018), https://www.forbes.com/sites/christopherelliott/2018/11/21/why-you-should-not-trust-online-reviews/?sh=5b003bae2218.

enough to address fake reviews on their sites."[538] In the United States, Amazon is trying to position its ban on certain sellers tied to a plethora of fake reviews as a voluntary measure, but internal memos show that the company is being pressured by the Federal Trade Commission (FTC) to crack down on the phonies.[539]

The world of Instagram, which is owned by Facebook, and the influencers it supports can be even more inauthentic. And the stakes just as high, or even higher. "Influencers on Instagram who have over a million followers can make more than $250,000 per post from brands, while someone like Kylie Jenner can make around $1 million for a single sponsored Instagram post."[540] For the 2021 HBO documentary *Fake Famous*, writer-director Nick Bilton conducted an experiment: could he pluck a few random people out of obscurity and turn them into Instagram stars? As it turns out, by doing things like buying social media followers—"$119.60 to buy about 7,500 followers and 2,500 likes"[541]—and renting a private jet studio for $49.99 an hour for a photo shoot,[542] he could.

While it might be difficult for the average viewer to spot a photo shoot that has taken place in a studio versus in an actual private jet, there are ways to tell the difference between, say, an Amazon review that has been written by a legitimately satisfied consumer, versus one that's been solicited and often, in some way, paid for by the seller. Services such as the

---

[538] Adria Calatayud and Joe Hoppe, "Amazon, Google Probed In U.K. Over Fake Reviews," *Wall Street Journal* (June 25, 2021), https://www.wsj.com/articles/amazon-google-probed-in-u-k-over-fake-reviews-11624615801?st=4x5udvqd-v3gz5xs&reflink=article_email_share.

[539] Jason Del Ray, "Amazon internal messages show the FTC is prodding the tech giant to punish fake-review schemers," Vox (May 20, 2021), https://www.vox.com/recode/22443153/amazon-seller-supsensions-aukey-mpow-ftc-paid-fake-reviews.

[540] Tom Huddleston, Jr., "How Instagram Influencers Can Fake Their Way to Online Fame," *CNBC* (February 2, 2021), https://www.cnbc.com/2021/02/02/hbo-fake-famous-how-instagram-influencers-.html.

[541] Ibid.

[542] Ibid.

aforementioned Fakespot and others, like ReviewMeta,[543] have developed algorithms that can help to weed out the fakes. All you, as a consumer, have to do is to plug the product link into their search bars, though there are some "tells" that you can spot for yourself. For example, reviews that are punctuated with multiple exclamation points in the text of the review, contain requests of the seller in the form of words or phrases such as "please," "offer more," or "bring back," or are posted by consumers who aren't verified buyers, should be suspect.[544] A most basic test, however, can be applied in this situation: if a review is too glowing, too enthusiastic, too fervent, then it might be one you're better off passing over.

For example, according to Cialdini, look at the number of 5-star reviews a product or service has received. If nearly all the reviewers have rated it with all five stars, something is likely fishy. "Of those review sites that we check before we make a purchase, the average number of stars that lead to a purchase is not five; it is a sweet spot of between 4.2 and 4.7 stars."[545] Anything more and your radar should start to remind you of that old rule of thumb: if something seems too good to be true, it probably is.

In the middle of the nineteenth century, this warning became popularized: believe nothing of what you hear and only half of what you see. In our age of Photoshop and digital enhancement, that warning takes on new layers of meaning, though the intent remains valid: question everything before you believe it and pass it off as fact. We no longer have an FDR, or even a Walter Cronkite-like figure, "the most trusted man in America,"[546] acting as the nation's faithful filter. The task now falls to us to keep from falling for foolishness.

---

[543] ReviewMeta website, https://reviewmeta.com.

[544] Eric T. Anderson and Duncan I. Simester, "Reviews Without a Purchase: Low Ratings, Loyal Customers, and Deception," *American Marketing Association* (2014), http://web.mit.edu/simester/Public/Papers/Deceptive_Reviews.pdf.

[545] Stephen J. Dubner, "How to Get Anyone to Do Anything," *Freakonomics* (May 26, 2021), https://freakonomics.com/podcast/frbc-robert-cialdini/.

[546] "Walter Cronkite Dies," *CBS* (July 17, 2009, https://www.cbsnews.com/news/walter-cronkite-dies/.

o o o

Finally, and in line with what we've talked about in the section directly above, we need to stay abreast of developments in the areas of tech, and our rights as they relate to it. Innovation happens fast in the tech world—*that* is just an indisputable fact. In order to make a decision about which new tech developments are going to be beneficial and worth integrating into our lives, we need to know what's coming. In our next and final chapter, we'll take a little tour of future tech to provide a baseline understanding of the direction tech is heading.

It may, at first blush, seem like a great deal of work to stay informed about the ever-changing, ever-innovating world of tech. A good grounding point is to keep in mind that your relationship with tech and how you choose to use it is, in many ways, a very personal one. As I'm sure you do, too, I know any number of people who seem obsessed with their technology and are linked in to a device and some other digital service nearly every waking hour—and I also know some folks who truly qualify as Luddites, only grudgingly using technology, and then only when they are absolutely forced to do so, if at all. We may judge their relationship with technology as healthy or unhealthy, depending on our own personal point of view, but it is, at the end of the day, *their* relationship. As with any relationship, if you want it to work for you, you have to work on it.

And you're better off if you don't go into it blind.

Chapter 7

Tomorrowland

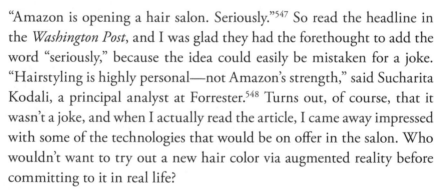

"Amazon is opening a hair salon. Seriously."[547] So read the headline in the *Washington Post*, and I was glad they had the forethought to add the word "seriously," because the idea could easily be mistaken for a joke. "Hairstyling is highly personal—not Amazon's strength," said Sucharita Kodali, a principal analyst at Forrester.[548] Turns out, of course, that it wasn't a joke, and when I actually read the article, I came away impressed with some of the technologies that would be on offer in the salon. Who wouldn't want to try out a new hair color via augmented reality before committing to it in real life?

Augmented reality (AR) is a technology that creates a composite view for a user by superimposing a computer-generated image over the user's view of the real world. Snapchat is poised to be first-to-market with an AR device—its 3D AR glasses are called, fittingly, "Spectacles"—beating

---

[547] Rachel Lerman, "Amazon is opening a hair salon. Seriously.", *Washington Post* (April 20, 2021), https://www.washingtonpost.com/technology/2021/04/20/amazon-hair-salon/.

[548] Ibid.

out bigger competitors like Facebook and Apple.[549] Indeed, in May 2021, Snapchat purchased its AR display supplier, WaveOptics, for $500 million.[550]

But what is *Amazon*'s end-game regarding its AR-equipped salon? To sell the technology to smaller stores, like your local salon? To create a platform upon which to sell more haircare products? According to Kodali, "It's throwing spaghetti against the wall. They're very experimental and they take what they can learn, and they will definitely have learnings from this—what technology people use, how do Amazon shoppers behave, what's the customer acquisition costs for a salon."[551]

At the other end of the spectrum, in the category of maybe-this-is-really-not-a-useful-development, I came across an article at Salon.com by new mother Julia Pelly, who was anxious about breast-feeding her newborn. In the article, she was taking to task the apps now available to new mothers regarding putting their newborns on a feeding schedule as "unnecessary and ineffective," a poor gap-filling measure in a culture that doesn't support breast-feeding mothers. The article's title? "Silicon Valley wants to 'optimize' your breast milk. Don't fall for it."[552] Or, in the same vein, what about the Snoo, the $1,500 robotic bassinet that uses AI and sensors to detect a fussy baby and soothe her, mimicking the conditions of the womb? Is it a godsend, facilitating much-needed sleep for infants and for parents, helping to prevent post-partum depression and Sudden Infant Death Syndrome by keeping babies strapped on their backs, which is the

---

[549] Sara Fischer, "Snapchat unveils its bid to dominate augmented reality," AXIOS (May 21, 2021), https://www.axios.com/snapchat-augmented-reality-ar-spectacles-snap-e824ccf4-6640-412e-a81a-6fec269aa9c0.html.

[550] Alex Health, "Snap Is Buying Its AR display Supplier for More Than $500 Million," *The Verge* (May 21, 2021), https://www.theverge.com/2021/5/21/22447150/snap-waveoptics-acquisition-500m-spectacles-waveguides.

[551] Ibid.

[552] Julia Pelly, "Silicon Valley Wants to 'Optimize' Your Breastmilk. Don't Fall for It," *Salon* (June 6, 2021), https://www.salon.com/2021/06/06/silicon-valley-wants-to-optimize-your-breastmilk-dont-fall-for-it/.

safest sleep position for infants—or is it "some ungodly combination of Mary Poppins and the HAL 9000"?[553]

In a conversation with Ralph Simon, founder of Mobilium Global, academic and author Vivek Wadhwa was asked the question: "How do we decide which technologies are to be accepted and which technologies are to be ignored?" His answer: "The technologies that uplift mankind are the ones we want to focus on." [554] That's just the sort of question we're going to have to ask ourselves more frequently than we might imagine now, as the advances in technology roll forward and into our lives with increasing rapidity. And Wadhwa's answer is an excellent standard to which we should aspire.

But the great majority of us aren't tech ethicists. We aren't the ones charged with examining technologies to ensure they meet ethical standards, that they don't exploit users' vulnerabilities or infringe on tech consumers' rights. Understanding the nuances of how a particular technology works, and following its operation and application to come to a conclusion about how it will uplift humankind or harm us, is beyond the scope of the day jobs most of us hold. In theory, the idea of being able to check out how you might look with a different hair color via augmented reality sounds like fun, but is it worth the potential risk of giving up even more of our personal information to the salon operator? Breastfeeding apps might be a godsend for some new mothers, but for others they may exacerbate the unbearably stressful situation of being unable to adequately nourish a child with one's own breast milk.

The fortunate development is that Big Tech companies seem to be taking ethics into serious consideration as their technology forges ahead. In late spring 2021, Google announced plans to double the size of its team studying the ethics of artificial intelligence in the coming years,

[553] Maura Judkis, "The Snoo is a $1,500 bassinet—and a touchstone for parental judgment, anxiety and privilege," *Washington Post* (July 13, 2021), https://www.washingtonpost.com/lifestyle/style/snoo-millennial-parents/2021/07/12/e9fa501a-e02e-11eb-9f54-7eee10b5fcd2_story.html.

[554] Vivek Wadhwa (Academic & Author) in Conversation with Ralph Simon," *YouTube* (June 9, 2020), https://www.youtube.com/watch?v=Q_ubGgsIEq8.

bringing the total number of researchers on the team to two hundred.[555] In the same timeframe, Facebook announced that over the course of the previous year, it had "quietly built up a small 'responsible innovation' team with a huge mission: Look at upcoming products and help prevent them from being misused to cause harm."[556]

Tech companies bringing in experts to evaluate the effects of innovations on the human community is a hopeful sign. Still, we have to keep in mind that we're dealing in large part with legislators who are so out of touch with technology and the culture around it that in order to focus their attention, they often decide upon those regulations by relying on what polling data tells them *we* are concerned about.

"Is Twitter the same as what you do?" Senator Lindsey Graham (R-SC) once asked Mark Zuckerberg, in an attempt to find out if Facebook was a monopoly.[557]

"How do you sustain a business model in which users don't pay for your service?" This question was again posed to Zuckerberg by Senator Orrin Hatch (R-UT). "Senator, we run ads," Zuckerberg replied. With a straight face.[558]

"What was Facemash, and is it still up and running?" Representative Billy Long (R-MO) asked that question, "much to Zuckerberg's embarrassment. If you've watched *The Social Network*, as Long evidently has, you know Facemash was an early Zuckerberg project in which users compared two photos of women and picked who was hotter. But Zuckerberg

[555] Tripp Mickle, "Google Plans to Double AI Ethics Research Staff," *The Wall Street Journal* (May 11, 2021), https://www.wsj.com/articles/google-plans-to-double-ai-ethics-research-staff-11620749048?&mod=djemfoe.

[556] Ina Fried, "Facebook's new plan to nip product misuse in the bud," AXIOS (April 7, 2021), https://www.axios.com/newsletters/axios-login-0e0600b5-cc8b-4fb0-9f8b-c36edfbf11ee.html?utm_source=newsletter&utm_medium=email&utm_campaign=newsletter_axioslogin&stream=top.

[557] Minda Zetlin, "The 9 Weirdest and Most Hilarious Questions Congress Asked Mark Zuckerberg," *Inc.*, https://www.inc.com/minda-zetlin/mark-zuckerberg-congress-hearings-funny-stupid-questions.html.

[558] Ibid.

started Facemash from his dorm room [in 2003], and Harvard shut it down within days."[559]

These are just a few of the more absurd questions that have been asked of tech insiders by members of the United States Congress, and, yes, our initial reaction is to have a good laugh at them. Shira Ovide, however, warns us that with laughter—dismissing legislators like these as "too old or too clueless to exercise effective oversight of tech superpowers"[560]—we are missing an important point: "tech companies are built around software that is designed not to be understood by outsiders."[561] She's right—and underscores the urgency of the calls I've made in this book for such basics as regulation to assure transparency from tech, and technology ambassadors to help design those regulations and put them into place. As one more example of the ill-informed manner in which our legislators are approaching rapidly emerging and changing tech, we can look to the way TikTok is being treated in Congress. TikTok is a social media platform used, in the main, by young people.[562] Indeed, TikTok has, as of October 2021, an astounding 1 billion monthly users. But "Recent Senate hearings—convened under the banner of 'Protecting Kids Online'—focused on a whistleblower's revelations regarding what Facebook itself knows about how its products harm teen users' mental health. That's an important question to ask. But if there's going to be a reckoning around social media's role in society, and in particular its effects on teens, shouldn't lawmakers also talk about, um, the platforms

[559] Ibid.

[560] Shira Ovide, "Congress Doesn't Get Big Tech. By Design.", *New York Times* (July 29, 2020), https://www.nytimes.com/2020/07/29/technology/congress-big-tech.html.

[561] Ibid.

[562] Katherine Schaeffer, "7 Facts About Americans and Instagram," *Pew Research Center* (October 7https://www.pewresearch.org/fact-tank/2021/10/07/7-facts-about-americans-and-instagram/.

teens actually use?"[563] And *that* is a critical question, because it is teens that use TikTok; "Facebook just *isn't cool*."[564]

The dangers teens face on TikTok, however, are many of the same ones they, and their parents, have to deal with on Facebook—exposure to body shaming, bullying, inappropriate and sexualized content.[565] Perhaps even more alarming are dangers unique to TikTok, such as the plethora of TikTok trends. Let me outline a few of them. There is the coronavirus challenge, in which people are encouraged to lick items in public—doorknobs and such; after a year of this challenge the hashtag #coronavirus-challenge had "3.1 billion views".[566] There's the Vampire Fangs challenge, which involves people using nail glue or superglue to affix fake fangs to their real teeth.[567] There's the Nutmeg challenge in which people are told to mix tablespoons of nutmeg with milk and drink it; the drink can cause intoxication—and it can also cause hypothermia, hallucinations, coma, and death.[568] You get the idea.

If our lawmakers are going to hold hearings that ostensibly focus on the online safety of our children, then they need, metaphorically speaking, to be playing in the same sandbox as our kids—or, at least, they need to know what sandbox our kids are playing in. The way that legislators are currently understanding social media platforms, their use, and by whom, is akin to trying to understand how a NASA rocket works by building a Lego™ rocket ship kit; you end up with rockets in both cases, but only one will get you to the moon. We'd consider a legislator who approached our

---

[563] Evelyn Douek, "1 Billion TikTok Users Understand What Congress Doesn't," *The Atlantic* (October 10, 2021), https://www.theatlantic.com/ideas/archive/2021/10/problem-underestimating-tiktok/620354/.

[564] Ibid.

[565] "How to Protect Teens from TikTok Dangers," *Global Learning* (September 26, 2019), https://globallearning.world.edu/2019/09/26/how-to-protect-teens-from-tiktok-dangers/.

[566] Seren Morris, "21 Dangerous TikTok Trends Every Parent Should Be Aware of," *Newsweek* (March 6, 2021), https://www.newsweek.com/21-dangerous-tiktok-trends-that-have-gone-viral-1573734.

[567] Ibid.

[568] Ibid.

space program in the latter way as having abdicated his or her responsibility, wouldn't we? We abdicate responsibility for the direction tech takes in the future—and we allow our legislators to abdicate too—at our peril.

That warning issued, however—and we will delve more deeply into the real dangers technology poses for us in the years to come before we're through—it is also a really exciting time in tech. Think of it: someone, somewhere, is right now dreaming up the next world-changing technology or technology-related breakthrough. Perhaps that breakthrough will be around an important piece of progress we've already talked about in this book. Perhaps our understanding of the fundamental fairness of being able to decide whether or not a thing as valuable and intimate as our personal information is a commodity, and whether or not to share it—and to be compensated for sharing it, if we decide to do that—will evolve, and tech users will reject the status quo of giving it away for free. Perhaps someone is right now taking the mathematical equation I've detailed in this book to calculate the value of an individual's 'net worth, and is working out both the legislation and the logistics it will take to make sure every individual who uses the internet receives a data dividend deposit into his or her bank account every month. Perhaps data dividends will become the new norm.

Or, perhaps, it will be something a little more dramatic, like the ability to have Whitney Houston or Maria Callas perform for you in your very own living room.

A hologram is a three-dimensional image formed by the interference of light beams from a laser or another coherent light source—a concept many of us are familiar with from the famous hologram scene in the original *Star Wars* movie, when a miniature Princess Leia appeals to Obi Wan Kenobi for help.[569] Holograms, however, can also be life-sized, and there are companies working now to bring full-blown concerts by the likes of Whitney Houston, Maria Callas, Roy Orbison, and Buddy

---

[569] "Princess Leia's Jologram Message," *YouTube*, https://www.youtube.com/watch?v=8N_Cj3ZS9-A.

Holly to a stage near you.[570] Currently, the technology is suitable only for large venues that can accommodate the equipment necessary to project the image—and, even then, there are limitations, in that certain seats can't be sold depending on projection angles. But someday it seems imaginable that we'll be able to dial up a holographic celebrity to perform a concert for us, right in our own homes. Indeed, holographic technology is progressing at such a pace that some experts are predicting that within the next decade, Facetime and Zoom calls will be replaced by live conversations with interactive holographic images of friends and family[571]—though holographic business meetings are already a thing. On attending a meeting in Shanghai as a holographic image from his office in Switzerland, Christoph Grainger-Herr, chief executive of the Swiss watch brand IWC remarked: "It's remarkably different from a Zoom call, because you're full-body, head to toe—and everybody is wearing pants."[572]

Then again, the ways in which we entertain ourselves may be undergoing a seismic and fundamental generational shift. Younger generations prefer very different forms of entertainment than older ones. For example, 18 percent of Millennials—those who were born between 1983 and 1996—chose watching TV shows and movies as their preferred form of home entertainment, but only 10 percent of members of Gen Z—people born between 1997 and 2007—chose this as their favorite pastime. Rather, 26 percent of Gen Z would rather play video games, 14 percent

---

[570] KC Ifeanyi, "The Hologram Concert Revolution Is Here," *Fast Company* (June 19, 2019), https://www.fastcompany.com/90365452/hologram-concert-revolution-like-it-or-not-meet-company-touring-whitney-houston-buddy-holly.

[571] Matt Clemenson, "We'll All Be Video-Chatting Each Other Over Hologram by 2030, Experts Predict," *Daily Star* (April 15, 2021), https://www.dailystar.co.uk/news/weird-news/well-video-chatting-each-over-23918124.

[572] Victoria Gomelsky, "Holograms: A Way to Be There in Spirit, if Not in Body," *New York Times* (June 18, 2021), https://www.nytimes.com/2021/06/18/fashion/watches-holograms-IWC.html?referringSource=articleShare.

would rather listen to music, and 21 percent said they'd rather browse the internet or interact on social media than watch a movie.[573]

What this all means for Jeff Bezos's 2021 purchase of the MGM catalog for $8.45 billion is yet to be determined.[574] What *will* it mean for Amazon to have a trove of so much classic content that dense? Well, we do know that the pandemic has inspired "several traditional media companies [to dive] into the streaming space, taking advantage of their vast video libraries and launching just in time to provide consumers in lockdown with more binge-worthy content."[575] As noted previously, subscriptions to streaming services doubled during the COVID-19 crisis. The Nielsen Corporation, a pioneer in data collection—in business since 1923, it is one of the world's first data-brokers[576]—quantified actual viewing in the Gauge, its monthly total TV and streaming snapshot. It "shows that [in fifteen months of lockdown] streaming usage across *all* television homes has climbed to 26% of all time spent on TV."[577] This gain was made, bear in mind, in the midst of the ongoing debate over which is more desirable, cable or satellite service or streaming service? That is, is it worth often paying double the monthly price tag for cable[578] over streaming, in order to avoid the service inter-

---

[573] Ryan Faughnder, "What entertainment does Gen Z prefer? The answer isn't good for Hollywood," *Los Angeles Times* (April 18, 2021), https://www.la-times.com/entertainment-arts/business/story/2021-04-18/what-entertainment-does-gen-z-prefer-the-answer-isnt-good-for-hollywood.

[574] Joseph Pisani, "Amazon To Buy MGM For $8.45 Billion," *Huffington Post* (May 26, 2021), https://www.huffpost.com/entry/amazon-to-buy-mgm_n_60ae43f6e4b03135479eb90f.

[575] "The Gauge Shows Streaming is Taking a Seat at the Table," *Nielsen* (June 17, 2021), https://www.nielsen.com/us/en/insights/article/2021/the-gauge-shows-streaming-takes-a-seat-at-the-table/.

[576] Nielsen website, https://sites.nielsen.com/timelines/our-history/.

[577] Ibid.

[578] Michael Timmermann, "Cable vs. Streaming: Does Cutting the Cord Really Save You Money?" *Clark* (June 30, 2020), https://clark.com/technology/tv-satellite-cable/cable-streaming-price-comparison/#:~:text=The%2520chart%-2520below%2520illustrates%2520the%2520differences%2520between%2520a,%2520%2520Yes%2520%25207%2520more%2520rows%2520.

ruptions that can and frequently do frustrate streaming service viewers, because "a platform's technology may be working with multiple third-party content delivery networks, authentication systems, different internet service providers and a plethora of devices that viewers are using to access their content"?[579]

This still leaves us with the question of what business model television viewing services will adopt in the years to come[580]—that is, how will they be monetized? Axios reports that "free, ad-supported apps have become hot acquisition targets for TV companies that want to sell digital TV ads, but don't have the digital TV audiences, and therefore digital ad inventory to do so."[581] In tandem with the development of

---

[579] J. Clara Chan, "Why Are Streaming Platforms Crashing Before Popular Finales?" *Hollywood Reporter* (June 18, 2021), https://www.hollywoodreporter.com/business/digital/hbo-max-mare-of-easttown-disney-plus-wandavision-showtime-logan-paul-floyd-mayweather-streaming-outages-1234969382/.

[580] The offerings in this category are heavy on acronyms: FASTS are free, ad-supported streaming services and include Disney's Hulu, Comcast's Xumo, ViacomCBS's Pluto TV, and Fox's Tubi, acquired in 2020 and already on its way to "becoming a $1 billion yearly revenue-driver" for the company; FNAS are free network apps that, because they are free, they frequently don't offer full shows to viewers but only ad-supported clips; vMVPDs, which are Virtual Multichannel Video Programming Distributors, often referred to as "skinny bundles," that aggregate live and on-demand TV but deliver the content over the internet, and may prove to be most attractive to current OTT users. (And *that* acronym, OTT, stands for Over-the-Top, a term used to describe the delivery method for TV as well as film content over the internet without the use of traditional broadcast, cable, or satellite TV providers. If, for example, you're paying for a high-speed internet connection through, say, Xfinity, and you use this connection to watch content on a service such as Netflix on your computer or your tablet but you're not paying for a cable connection, you're an OTT consumer.); finally, we have the hybrid services such as CBS All Access, which use, indeed, a hybrid business model, being both paid subscription and ad-supported. Subscribers pay a smaller fee than those who opt for an ad-free service, but are served fewer ads than those who pay no subscription fee at all.

[581] Sara Fischer, "Free apps are driving adoption of digital TV ads," AXIOS (March 12, 2019), https://www.axios.com/free-apps-adoption-digital-tv-ads-b0a58d7d-90f8-4ee7-b4fa-e87b3bcceec4.html.

new business models, we have, additionally, the Scott Galloway-coined concept of "rundling."

Bundling has long been a staple part of retail sales strategies. Both brick-and-mortar and online retailers frequently pull together multiple products to sell as kits or more complete units—from electronics retailers who build a stereo system from the most-desired component parts and sell the whole as one discounted package, to the manufacturers of haircare products who have made bundling into an art, packaging shampoo and conditioner together in one low-priced box. Even grocers get in on the game: buy one top round roast and get the second for half off. Consumers like bundles because it saves them the time, headache, and money that might have otherwise gone into comparison shopping in order to create a package for themselves. And it is good for businesses, in that bundling increases the value of an average sale without increasing the customer acquisition cost, and it allows the retailer to both introduce lesser-known products and to offload overstocks to the consumer.

The 'r' in rundling represents the concept of *recurring revenue*. For example, it is nice for a retailer to sell you one blouse; it is better if they sell you a blouse every month. This is the model that a company such as Stitch Fix has adopted: curating monthly fashion boxes with pieces that suit your individual style, items that you can try on at home and then keep or return as you see fit. The company charges a monthly twenty-dollar "styling fee [to cover] your stylist's expertise and time—and it gets credited toward anything you keep."[582] The model works across industries, of course. Take Adobe, which, in 2012, stopped selling its creative and design tools as perpetual-license software for hundreds of dollars a set, and moved to a lower-priced subscription model. As a result, Adobe's net earnings fell for the next consecutive three years; but the CFO Mark Garrett kept faith in the new model and, by the end of 2015, "Adobe achieved revenue of $1.11 billion, above the high end of the targeted range

---

[582] Stitch Fix website, https://www.stitchfix.com.

of $1.05 billion to $1.10 billion.[583] Indeed, Adobe's "earnings have grown faster (45.44% per year) than the U.S. Software-Infrastructure industry average (31.13%)."[584]

Rundling, then, is the concept of selling a bundle of services to you for one fixed, low, recurring monthly fee. Imagine how rundling could impact the business of streaming, offering customers the binge power of getting Netflix ($8.99/month),[585] Hulu ($6.99/month), NBC's Peacock Premium ($4.99/month), Amazon Prime Video ($8.99/month), Disney+ ($7.99/month), HBO Max ($9.99/month), YouTube TV ($64.99/month), Magellan TV ($6.99/month), PBS Documentaries ($3.99/month), Discovery+ ($4.99/month), fuboTV ($64.99/month), ESPN+ ($6.99/month),[586] all for one low price—say $24.99 versus the current combined market price of $200.88 a month to buy each service separately—plus the convenience of one monthly automated payment instead of ten or twelve or fourteen. As Galloway forecasts: "Business has mistaken 'choice' as a good thing. Consumers want less choice, but instead, confidence in the (fewer) choices presented to them. We begin to witness a finite series of consumer networks: media, apparel, travel, and health. It's going to be an arms race, with cheap capital as the munitions, to see what brands and retailers can establish credible/compelling bundles."[587]

---

[583] "Adobe Reports Strong Q1 FY2015 Financial Results," *Adobe* website (March 17, 2015), https://www.businesswire.com/news/home/20150317006384/en/Adobe-Reports-Strong-Q1-FY2015-Financial-Results#:~:text=Adobe%20achieved%20revenue%20of%20%241.11%20billion%2C%20above%20the,new%20subscription%20additions%20in%20Q1%20fiscal%20year%202014.

[584] "ABDE Past Earnings Growth" *WallStreetZen,* https://www.wallstreetzen.com/stocks/us/nasdaq/adbe/earnings.

[585] All prices indicated in this paragraph are current basic prices as of July 2021.

[586] All prices cited are lowest prices available for basic service as of October 14, 2021.

[587] Anne Gherini, "The Rise Of The Rundle: A New Trend For Subscription-Based Services," *Inc.* (February 28, 2019), https://www.inc.com/anne-gherini/the-rise-of-rundle-a-new-trend-for-subscription-based-services.html.

Even public media—such as the Public Broadcasting System (PBS), funded through the Corporation for Public Broadcasting (CPB), that has for decades brought us everything from the PBS NewsHour to Sesame Street—is considering a revamp to remain relevant in our increasingly digital culture. "There absolutely has to be a much bigger role for non-profit media, with public media as a subset of that, than there has been in the past," said Steve Waldman, president of Report for America.[588] "Right now a disproportionate amount of CPB money goes to TV," Waldman added. "From a local news point of view, we may need to loosen that up and have the money go to wherever it can strengthen local programming."[589] The revamp, detailed in a report from the German Marshall Fund of the United States considers "what a reinvigorated infrastructure for civic information might look like,"[590] and how that involves not only—or not disproportionately—broadcast television as it has in the past but digital platforms, as well as more diverse content producers including educational institutions, local governments, nonprofit organizations, and independent journalists.

As we're talking about all the ways we're going to inform and entertain ourselves in the years and decades to come, let's not forget that our youngest generations have indicated that they would prefer to play video games rather than watch a television show or movie in their leisure time. Indeed, the video game industry is worth more than $120 billion—more than movies and music combined—and is expected to reach a worth of $200 billion by 2022.[591] Think of Twitch's thirty mil-

[588] Kim Hart, "The push for a 'PBS for the internet,'" AXIOS (August 2, 2021), https://www.axios.com/pbs-internet-online-information-nonprofit-e9a78344-ae58-4214-9cb6-974ab9576cb1.html.

[589] Ibid.

[590] Sanjay Jolly and Ellen P. Goodman, "A 'Full-Stack' Approach to Public Media in the United States," *The German Marshall Fund of the United States* (July 2021), https://www.gmfus.org/sites/default/files/Jolly%2520and%2520Goodman%2520-%2520public%2520media.pdf.

[591] Stephen Johnson, "3 Ways Video Games Will Evolve in the 2020s," *Big Think* (January 14, 2020), https://bigthink.com/technology-innovation/future-video-games?rebelltitem=1#rebelltitem1.

lion users a day as another way to quantify the startlingly rapid growth of the industry. Some of this growth can be attributed directly to the sort of data collection we've talked about. "Today, the shift to online and mobile gaming has given game developers and publishers the ability to collect and process massive quantities of data about players: not only what they like to play, but what they read online, who they play with, and what makes them spend money."[592] That is, simply by playing the games, gamers provide developers with the very information they need to make the games ever more compelling—*addictive*—to their players, and more lucrative for themselves.

Even so, make no mistake: we can expect continued growth in this industry, some of it centered around improvements in the execution of video games themselves. Advances in ray tracing technology,[593] for example, which enhances the realistic look within the games; and developments in AI, specifically "motivated reinforcement learning,"[594] which is poised to allow currently clunky non-player characters (NPC) to react to and learn from the game environment and their experiences within it. That will likely only increase the audience for this form of entertainment. On the other hand, the younger generations might decide, at least from time to time, to go old-school with their games—or, anyway, *relatively* old-school—facilitated by inventions such as the Infinity Game Table, a touchscreen tabletop PC dedicated to games like Monopoly, Scrabble, Sorry, and even Hungry Hungry Hippos.[595]

---

[592] Joshua Foust and Joseph Jerome, "A Guide to Reining in Data-Driven Video Game Design," *Brookings Tech Stream* (June 25, 2021) https://www.brookings.edu/techstream/a-guide-to-reining-in-data-driven-video-game-design-privacy/.

[593] "Technology Sneak Peek: Advances in Real-Time Ray Tracing," *Unreal Engine* (November 20, 2018), https://www.unrealengine.com/en-US/tech-blog/technology-sneak-peek-advances-in-real-time-ray-tracing.

[594] Kathryn Merrick and Mary Lou Maher, "Motivated Reinforcement Learning for Non-Player Characters in Persistent Computer Game Worlds," *ACM Digital Library* (June 14, 2006), https://dl.acm.org/doi/10.1145/1178823.1178828.

[595] Ina Fried, "Infinity Game Table Makes Game Night Come Alive," AXIOS (June 23, 2021), https://www.axios.com/infinity-game-table-makes-game-night-come-alive-a3d2201c-8f6d-4b83-8f16-396b4d19e44a.html.

o o o

Now, while it's going to be satisfying to see the balance bump when our data dividends arrive in our bank accounts each month, and truly exhilarating to watch how performers will transform their touring models as they begin to experiment with using holograms to entertain us, we need to consider what else is afoot in the techno realm—and then be prepared to be realistic, even hard-nosed, about the changes that technology innovators have in store for us. Some of those changes are going to be disappointing to some, and others will be dangerous, even terrifying, for the world at large; but they are all coming, and more in the near term than we might imagine.

Let's start with changes in the area of digital advertising, which are likely to have big repercussions on the way we use social media, too, particularly regarding Facebook. A common bit of now-conventional wisdom for those who advertise digitally is that "organic is dead." This means that a business hoping to get the word out about a product or service can't rely on reaching even the people who have already liked or followed his business. Or, to put it another way, just because you have expressed interest in your town's new greenhouse/nursery doesn't mean that Facebook is going to serve you that nursery's posts in your newsfeed *unless that nursery has paid either for an ad or to "boost" its posts.* The posts Facebook's algorithms are going to "choose" to serve you, once they have learned that gardening is your hobby, will be about gardening, of course, but they will in large part be paid posts. We all recognize that another bit of wisdom about life itself is that nothing lasts forever—old things inevitably fade and give way to new. Facebook may well become more of an advertising and marketing platform in an algorithmically organic way, and less of a social media site, neatly setting itself up to be replaced by the next big social media thing.

Contrast what is, in the grand scheme of things, the relatively minor transformation of Facebook to quite another transformation that is much more profound, and dangerous.

Eric Schmidt, former Google CEO and Chair of the National Security Commission on Artificial Intelligence warns, "For the first time since World War II, America's technological predominance—the backbone of its economic and military power—is under threat."[596] The warning continues in a 756-page report issued in March, 2021: "China possesses the might, talent, and ambition to surpass the United States as the world's leader in AI in the next decade if current trends do not change. Simultaneously, AI is deepening the threat posed by cyberattacks and disinformation campaigns that Russia, China, and others are using to infiltrate our society, steal our data, and interfere in our democracy."[597] For example, the ownership of patents is "widely seen as an important sign of a country's economic strength and industrial know-how."[598] In 2020, China edged out the United States from the top spot it has held since the international patent system was sent up forty years ago, filing 58,990 patent applications to the United States's 57,840.[599] A year earlier, in 2019, China had already beaten the United States in the number of patents secured by academic institutions for innovations in AI technology.[600] Schmidt warns, "We're in a geo-political strategic conflict with China. The way to win is to marshal our resources together to have national and global strategies for the democracies to win in AI. If we don't, we'll be looking at a future where other values will be imposed on us."[601]

As one of the world's super powers, as well as the world's essential democracy, the United States may be ceding its role in other ways as well, making way for the next big thing in world power—and in cyberwarfare. In the past, for example, meetings between American presidents and Russian ones have been "dominated by one looming threat: the vast nuclear arsenals that the two nations started amassing in the 1940s, as

---

[596] "Final Report," *National Security Commission on Artificial Intelligence* (2021), https://www.nscai.gov/wp-content/uploads/2021/03/Full-Report-Digital-1.pdf.

[597] Ibid.

[598] Stephanie Nebehay, "In a First, China Knocks U.S. from Top Spot in Global Patent Race," *Reuters* (April 7, 2020), https://www.reuters.com/article/us-usa-china-patents-idUSKBN21P1P9.

[599] Ibid.

[600] "Microsoft President: Orwell's 1984 Could Happen in 2024," *BBC* (May 27), https://www.bbc.com/news/technology-57122120.

[601] Ibid.

instruments of intimidation and, if deterrence failed, mutual annihilation."[602] But at the June 2021 meeting in Geneva between President Biden and Russian president Vladimir Putin, "for the first time cyberweapons [were] elevated to the top of the agenda."[603] Let me remind you that while Putin's "arsenal of cyberweapons…is put to work every day,"[604] the United States remains a good twenty-five years behind not only in its legislative answers to tech dominance, but also in its conceptualization of what it means to live in an increasingly tech-centered world, even when that world is at peace. Perhaps even more unnerving than the United States' lack of preparedness to meet the technological future is that private concerns are stepping in to fill that gap. Palmer Luckey, "best known for selling Oculus to Facebook in 2014 for $2 billion before he was fired in 2017 amid controversy for his political donations and financial support of far-right groups," announced in June of 2021 that his start-up, Anduril, with which he hopes to "turn allied warfighters into invincible technomancers," is now valued at $4.6 billion.[605]

Does the future include a privatized army? And if our armed forces are privatized, to whom then does the power—or ability—to declare war fall? If the support and sustenance of our troops is taken out of the hands of the Congress, does it follow that the power to deploy them can, as well?

As we have already seen—from interference in U.S. elections to shutting down our oil pipeline and meat production facilities—warfare in the coming decades is going to be waged in wholly unprecedented ways. And, potentially, by an armed entity wholly and materially different from the Army, Navy, Air Force, and Marines as we know them today. The experts

---

[602] David E. Sanger, "Once, Superpower Summits Were About Nukes. Now, It's Cyberweapons.", *New York Times* (June 15, 2021), https://www.ny-times.com/2021/06/15/world/europe/biden-putin-cyberweapons.html?referringSource=articleShare.

[603] Ibid.

[604] Ibid.

[605] Todd Haselton, "Ex-Facebook VR Exec Says He'll Turn U.S. Troops into 'Invincible Technomancers,' Just Raised $450 Million," *CNBC* (June 21, 2021), https://www.cnbc.com/2021/06/17/anduril-turning-us-troops-into-invincible-technomancers-palmer-luckey-says.html.

are being very clear that the United States is not ready to compete on this new field of battle. Take this as "a clear indication of how bad the cyber threat has become".[606] if you visit the website of the National Security Agency, "you'll come to a link for what the NSA calls its 'Cybersecurity Collaboration Center' for sharing ideas with tech companies about stopping malware attacks."[607] That the NSA is, essentially, crowdsourcing national security is, well, *concerning*.

Importantly, some journalists are concerned the media aren't prepared to present the case to us, anyway, because "one of the problems is that there simply aren't very many journalists that fully understand neither how cyberattacks work nor what they are."[608] As we scan our daily newspapers and catch up on the Sunday news shows—and as the media decide on their content—we need to weigh what it is we're going to worry about, and what it is we need to fix.

o o o

From self-driving cars that are, essentially, digital devices on wheels, to planes that can take you from New York to Paris in time for lunch and bring you back again in time for dinner, travel is going to look a whole lot different in the future. Think I'm kidding about the planes? That I'm just indulging in a Jetsons-style futuristic fantasy? Scientists in the U.K. and China have developed a special coating for planes, made of zirconium carbide, that could make hypersonic air travel possible.

A major impediment to practical hypersonic travel in the past has been that, at speeds of Mach 5 or even greater—which means planes that can travel at speeds at least five times the speed of sound—heat builds up

---

[606] David Ignatius, "An undeclared war is breaking out in cyberspace. The Biden administration is fighting back", *Washington Post* (August 10, 2021), https://www.washingtonpost.com/opinions/2021/08/10/an-undeclared-war-is-breaking-out-cyberspace-biden-administration-is-fighting-back/.

[607] Ibid.

[608] Cyrus Farivar, "A Brief Examination of Media Coverage of Cyberattacks (2007–Present), *Cryptology and Information Security Series* (Volume 3, 2009), https://www.arifyildirim.com/ilt510/cyrus.farivar.pdf.

on the aircraft and "temperatures can reach an insane 2,000 to 3,000°C. At such tremendous speeds, even the toughest plane currently faces oxidation and ablation as hot air and friction start to remove the surface layers of the plane."[609] The zirconium carbide coating, however, may mitigate that problem and actually allow for lunch in Paris and dinner in New York—and a late-night snack in Hawaii, if you're so inclined. We'll be presented with a new problem if—and *when*—such hyper-rapid transit becomes possible: how to deal with all the jet lag that will surely result. But without too much doubt, we can bet on science to come up with a cure for that, too.

One thing we can be certain will play a huge part in future tech is speed—the speed of a plane, and the speed of computers themselves. *Quantum computing*, according to Microsoft, which has a robust program to develop this capability, is "the use of quantum mechanics to run calculations on specialized hardware,"[610] which goes a long way toward explaining what quantum computing is to only a subset of actual engineers. "[Scientists] sent fifty indistinguishable single-mode squeezed states into a 100-mode ultralow-loss interferometer and sampled the output using 100 high-efficiency single-photon detectors. By obtaining up to 76-photon coincidence, yielding a state space dimension of about 1030, they measured a sampling rate that is about 1014-fold faster than using state-of-the-art classical simulation strategies and supercomputers,"[611] according to *Science Magazine.* That jargon actually helps to clear things up, because the point of quantum computing is that it is *fast*: Google has created a quantum computer that is *100 million times faster* than any

---

[609] David Grossman, "New Ceramic Plane Coating Could Be Used in Hypersonic Flight," *Popular Mechanics* (July 10, 2017), https://www.popularmechanics.com/technology/infrastructure/a27245/new-ceramic-coating-hypersonic-flight/.

[610] "What Is Quantum Computing?" *Azure*, https://azure.microsoft.com/en-us/overview/what-is-quantum-computing/.

[611] Han-Sen Zhong, Hui Wang, Jian-Wei Pan, et. al, "Quantum Computational Advantage Using Photons," *Science* (December 18, 2020), https://science.sciencemag.org/content/370/6523/1460.

other, classical computer.[612] A "classical computer," for the record, is any computer that is not a quantum computer—i.e., the sort that sits on top of most of our desks these days. Imagine, if you will, a computer in your office that makes the computations necessary to solve a problem 100 million times faster than the one you sit before every day. With the right input, a computer like that could likely solve the thorny problem of how to calculate an individual's 'net worth—and the amount of her monthly data dividend check—faster than a zirconium carbide-coated plane could get you from Chicago to Tokyo.

o o o

For all the damage tech has helped to inflict on our democracy—allowing the flow of disinformation to flourish, conspiracy theories to foment, and foreign interference to go unchecked in our elections—maybe, just maybe, tech can redeem itself as a useful democratic tool. There is at least one tech CEO, Tim Cook of Apple, who believes that we should be able to vote on our phones.[613]

While some experts believe that "the only safe election is a low-tech election,"[614] with paper ballots and possibly pebbles dropped into urns, the method used by ancient Greeks,[615] Cook thinks that tech-centric voting would actually be the answer to some modern voting issues, like fraud. "I would dream of that, because I think that's where we live. We do our banking on phones. We have our health data on phones. We have

---

[612] Google AI Blog website, https://ai.googleblog.com/2015/12/when-can-quantum-annealing-win.html.

[613] Ben Gilbert, "Tim Cook Wants Americans to Be Able to Vote on Their iPhones," *Insider* (April 5, 2021), https://www.businessinsider.com/apple-ceo-tim-cook-on-voting-technology-iphones-smartphones-2021-4.

[614] Kevin Roose, "The Only Safe Election is a Low-Tech Election," *New York Times* (February 4, 2020), https://www.nytimes.com/2020/02/04/technology/election-tech.html.

[615] Ibid.

more information on a phone about us than is in our houses. And so why not?"[616]

Considering the drawn-out resolution to the 2020 U.S. presidential race—and the even more drawn-out refusal of a portion of the population to accept valid election results amid accusations of non-existent voter fraud—proponents of voting by phone, or via the internet, aren't likely to find a wholly receptive audience among any faction of the American public in the near future. Still, experiments with digital voting continue,[617] and currently, thirty-two states allow some form of digital ballot-casting, in most cases under the auspices of the Uniformed and Overseas Citizens Absentee Voting Act (UOCAVA), which allows military service members to cast online ballots for local and national elections in accordance with their state's specific evoting laws.[618] For now, online voting continues to be "an academic research project," according to Dan Wallach, a computer science professor at Rice University.[619] But if security concerns can be adequately—meaning thoroughly—resolved, voting with one's phone or on one's computer could mitigate one of America's real ongoing and long-term election problems, and that is low voter turnout. Every chance we have to make it easier for citizens to vote, the closer we come to the more perfect democracy our Founders urged us to aspire to create. Maybe currently suspicious tech-centric voting methods might actually enhance democracy in the future.

---

[616] Gilbert, ibid.

[617] Miles Parks, "In 2020, Some Americans Will Vote On Their Phones. Is That The Future?" *NPR* (November 7, 2019), https://www.npr.org/2019/11/07/776403310/in-2020-some-americans-will-vote-on-their-phones-is-that-the-future.

[618] Mia Logan, "These States Allow Online Voting for Citizens, Is Your State One of Them"? *eBallot*, https://www.eballot.com/blog/these-states-allow-online-voting-for-their-citizens-is-your-state-one-of-them.

[619] Nathaniel Lee, "Here's Why Most American Are Not Able to Vote Online in 2020," *CNBC* (September 23, 2020), https://www.cnbc.com/2020/09/23/why-us-cant-vote-online-in-2020-presidential-election-trump-biden.html.

o o o

Okay, let's clear up a few things: there are no microchips being inserted under the skin of the recipients of any one of the COVID-19 vaccines; the vaccines do not alter your DNA; and you aren't magnetized after you receive one.[620]

But tech *can* change your body, and in the right circumstance, you may want it to, or *need* it to do just that.

We'll start out with the unwelcome body changes tech can cause, the ones we read about most and every computer user fears: the sitting-related ailments that more and more of us suffer from as the use of tech has become dominant in our lives. Sitting for hours at a stretch in front of our computers or gaming consoles can bring on ailments of two different varieties: musculoskeletal and metabolic. The musculoskeletal ones are caused because of the posture we assume when we sit—shoulders hunched, head pushed forward, pelvis tilted back. Sustaining this posture for four, six, eight or more hours a day can, over time, lead to chronic back and neck pain, and breathing problems. It can lead as well to rheumatic disorders, like arthritis, and over an even longer period of time may lead to evolutionary changes, in which future generations are, from birth, adapted to the tech lifestyle, with hunched backs and shorter arms.[621] Some experts warn that millennials are already dealing with "tech neck" and the sort of back problems usually reserved for people twice their age because of overuse of their smartphones, tablets, and other devices.[622]

---

[620] Jemima McEvoy, "Microchips, Magnets and Shedding: Here are 5 (Debunked) Covid Vaccine Conspiracy Theories Spreading Online," *Forbes* (June 3 2021), https://www.forbes.com/sites/jemimamcevoy/2021/06/03/microchips-and-shedding-here-are-5-debunked-covid-vaccine-conspiracy-theories-spreading-online/?sh=714ea9926af0.

[621] Francesca Specter, "'Office Worker of the Future' Has a Permanent Hunchback," *Yahoo News* (October 24, 2019), https://news.yahoo.com/office-worker-of-the-future-hunchback-074105101.html.

[622] Phoebe Weston, "Millennials Are Turning Into Hunchbacks," *Daily Mail* (March 6, 2018), https://www.dailymail.co.uk/sciencetech/article-5467557/Millennials-turning-HUNCHBACKS-warn-experts.html.

The metabolic problems are caused because when we sit at a computer or gaming console, we aren't burning anywhere near as much energy as the human body is designed to burn. The sedentary tech lifestyle can lead to general, overall muscle weakness, cardiovascular problems, trouble with blood pressure regulation, obesity, and even certain types of cancer. Indeed, some research has put the increased risk of a sedentary lifestyle leading to death from cancer at 82 percent.[623]

The way to combat these problems, according to experts, is to make sure your computer monitor sits at a 90-degree angle in front of your eyes when you're working at it, so you can maintain a better posture, and to never work with your laptop or tablet directly in your lap. Additionally, getting out from behind your device at least once an hour for five or ten minutes of stretching, or to take a brisk walk for a minimum of thirty minutes a day, can go a long way toward mitigating these physical issues. When I know I am going to be spending longer hours at my desk, among the tricks I've found helpful is to set an alarm on my phone to go off every hour and remind me it's time to take a short break.

Look, these are real health issues, and they're going to have a long-term impact not only on our own individual health and wellness, but on the health and longevity of our children, and their children. They're also going to have an impact on our personal healthcare costs and, down the road, on government budgets for healthcare. That is, they will unless we take them seriously now, and do the small things we need to do to adapt to living with tech in thoughtful and proactive ways.

○ ○ ○

On the flip side, looking at the benefits tech brings to our lives and health, we have a trove of possibilities. Let's start with the microchips that actually can be—and *are* being—implanted in the human body.

In 2013, in the tech-savvy country of Sweden, former professional body-piercer Jowan Österlund founded a company he called Biohax

---

[623] Sandee LaMotte, "Too Much Sitting Raises Your Risk for Cancer, Study Finds," *CNN* (June 18, 2020), https://www.cnn.com/2020/06/18/health/sitting-cancer-study-wellness/index.html.

International, and began to insert microchips into the hands of his fellow countrymen. The chips, about the size of a grain of rice, were inserted into the skin between the thumb and forefinger, and they allowed those who were chipped to do things such as unlock their home and office doors, share contact information, and board a train without a ticket, simply by a wave of their hand.[624] In 2017, Österlund brought his chips to the Three Square Market, a Wisconsin vending machine company, where U.S. employees lined up to have the chips implanted. The "chipping party" was televised—a marketing coup by the company, which was hoping to interest customers in being able to pay for vending machine purchases with a wave of their microchipped hands, and for Österlund, who was hoping to bring the chips to a global market.[625] The event was not without controversy, as you can imagine. Objections included concerns that the chips could be hacked, whether the privacy of those who'd been chipped was compromised, and even if a biblical end-times prophecy was being fulfilled because those employees now carried "the mark of the beast."

Even so, the demand for radio-frequency identification chips (RFID) is growing,[626] spurred on by the COVID-19 epidemic and Pentagon researchers who created a chip that could detect a coronavirus infection before the person began to show symptoms.[627] The RFID industry imagines uses for the implanted chips beyond unlocking the doors of

---

[624] Maddy Savage, "Thousands of Swedes Are Inserting Microchips Under Their Skin," *NPR* (October 22, 2018), https://www.npr.org/2018/10/22/658808705/thousands-of-swedes-are-inserting-microchips-under-their-skin.

[625] Oscar Schwartz, "The rise of microchipping: are we ready for technology to get under the skin?", *The Guardian* (November 8, 2019), https://www.theguardian.com/technology/2019/nov/08/the-rise-of-microchipping-are-we-ready-for-technology-to-get-under-the-skin.

[626] Marketing Insights Reports, November 8, 2021, https://www.openpr.com/news/2453487/uhf-rfid-chip-market-by-manufacturers-rising-demands.

[627] Michael Lee, "Pentagon Scientist Develop Microchip That Detects COVID Before Symptoms When Placed Under Skin," *MSN* (April 13, 20121), https://www.msn.com/en-us/news/technology/pentagon-scientists-develop-microchip-that-detects-covid-before-symptoms-when-placed-under-skin/ar-BB1fChwa.

the users' homes and offices, riding public transit, and paying for snacks with the wave of a hand. They envision such things as replacing tickets for concerts and sporting events, replacing credit cards, holding contact information for easy transmission, containing a person's whole healthcare history for more overall and efficient healthcare, and interfacing with our devices through a chip implanted in our hands or wrists rather than by way of a keyboard or touch system. Imagine walking into the office of your new doctor, and rather than being handed a clipboard, a pen, and a stack of paperwork to fill out, you simply wave your hand over a receiver and that information is transferred automatically into your record. That sort of ease will come, if the technology can be reconciled to eliminate privacy concerns now.

We are still discovering the countless ways in which technological advancements can enhance, improve, or augment our bodies. Among them are robotic exoskeletons—essentially wearable electromechanical devices, conceptually not unlike the suit of armor Tony Stark wears as Ironman. Perhaps surprisingly, the first suit of this nature was developed in 1890 by Russian engineer Nicholas Yagin, in an attempt to improve the mobility of people with physical impairments. The technology to power the suits has changed from compressed gas and steam in the concept's earliest days, to the electric motors, pneumatics, and hydraulics that go into contemporary suits. But one of the goals remains to make life-changing improvements in mobility for people who, because of injury, stroke, or even normal aging, might otherwise be confined to wheelchairs.[628] The potential of the suits doesn't stop there, however. "Exosuits" could be used by warehouse workers and delivery people to amplify their strength and protect back muscles while lifting heavy parcels and pallets. They could be used as protective gear in a host of professions, from lumberjack to firefighter to farm worker to combat soldier. "Overexertion by workers costs employers $15 billion a year in compensation," according to one

---

[628] Steven Ashley, "Robotic Exoskeletons Are Changing Lives in Surprising Ways," NBC (February 21, 2017), https://www.nbcnews.com/mach/innovation/robotic-exoskeletons-are-changing-lives-surprising-ways-n722676.

report by NBC News;[629] perhaps exosuits might initiate, in turn, many new and improved safety standards.

Bionic eyes are no longer reserved for fiction, either. Currently, cameras mounted to a pair of glasses and feeding input to electrodes attached directly to the retina are being used to treat macular degeneration, effectively reversing this form of blindness.[630] And X Lab—formerly Google X, under its Alphabet umbrella—is developing a smart contact lens that can be used by diabetics to monitor blood glucose levels in tears, and so alert wearers when their blood sugar is dipping too low.[631] *Bioprinting* is a real thing, too, with scientists and researchers working to 3D-print living human skin, nerve cells, hearts, kidneys, bone, and limbs. It's an endeavor that *PC Magazine* described as "creepy but incredible"[632] when they wrote about the work in 2015; and while that still may be an apt description, think of the possibilities. For instance, patients in need of, say, a heart transplant won't be beholden to waiting for tragedy—the death of another—to receive the organ they desperately need: compatible, patient-specific organs could be generated through a 3D printer on demand.

o o o

Perhaps the most astonishing branch of this category of science is *neurotech*, "a catchall term broadly encompassing an industry set on connecting human brains to computers."[633] To date, as far as clinical applications go, doctors are already having success in helping patients who suffer with

---

[629] Ibid.

[630] Bionic Vision Technologies website, https://bionicvis.com.

[631] Brian Otis and Babak Parviz, "Introducing Our Smart Contact Lens Project," *Google blog* (January 16, 2014), https://blog.google/alphabet/introducing-our-smart-contact-lens/.

[632] Chandra Steele, "The Creepy But Incredible World of 3D-Printed Body Parts, *PC Magazine* (July 17, 2015), https://www.pcmag.com/news/the-creepy-but-incredible-world-of-3d-printed-body-parts.

[633] Dalvin Brown, "Your tech devices want to read your brain. What could go wrong?", *Washington Post* (April 27, 2021), https://www.washingtonpost.com/technology/2021/04/27/brain-controlled-tech-facebook-neurable/.

Parkinson's Disease control their balance, tremors, and issues with walking by delivering low-intensity electrical pulses to nerves and surrounding tissues in the brain.[634] Ongoing research shows promise in the treatment of other neurological conditions, including ADHD, Alzheimer's disease, and epilepsy, as well as in relieving depression, anxiety, and even insomnia.

It is the leap that this branch of science makes from clinical applications that might prove most astounding—both exciting and off-putting. "There is nothing either good or bad, but thinking makes it so," William Shakespeare wrote in *Hamlet* (Act 2, Scene II) four hundred years ago. Contemporary scientists are working to demonstrate how prescient Shakespeare truly was: Facebook is among the companies "researching how to take minute nerve movements in your arm and translate them into gesture controls for your gadgets."[635]

How would that *work?* According to Nathalie Gayraud, a research scientist at Facebook Reality Labs, "the prototype looks like a thick black iPod strapped around the wrist. In theory, sensors on the device would be able to pick up what hand movements you intend to make through a technique for recording nerve signals known as electromyography (EMG). 'If you send a control to your muscle saying, "I want to move my finger," it starts in your brain. It goes down your spine through motor neurons, and this is an electrical signal. So you should be able to grab that electrical signal on the muscle and say, "Oh, okay. The user wants to move the finger.""'[636] In simple terms, this technology works through the thoughts we think every second of every day; it detects the electrical impulses in the body that are generated by our thoughts and acts on them…so we don't have to. It is technology that responds not just to a push of a button, but to our *intentions.*

---

[634] "Summary of Safety and Effectiveness Data," *Food & Drug Administration* (June 12, 2015), https://www.accessdata.fda.gov/cdrh_docs/pdf14/P140009b.pdf.

[635] Dalvin Brown, "Facebook reveals plan to let you control augmented reality with your thoughts," *Washington Post* (March 18, 2021), https://www.washingtonpost.com/technology/2021/03/18/facebook-wrist-control-thoughts/.

[636] Ibid.

Imagine merely thinking that you'd like to peruse your favorite morning newspaper, and the home page of the *Washington Post* appears on the screen of your iPad. Or reading in bed at night and merely thinking you'd like to turn on the lamp on the bedside table, in order to be bathed in light. Or sitting at your desk and merely thinking that you'd better preheat the oven to prepare dinner, and the oven is already on its way to 350 degrees. It won't be magic—we'll still have to get up from behind our desk to actually make dinner, as there are no plans in the offing for a brain/computer interface that chops the onions—but it might just feel like it. At least, it will to those of us who will still remember what it was like to have to go into the kitchen to turn on the oven.

o o o

Advances in the realm of virtual reality may be just as stunning. *Haptic technology,* also known as kinaesthetic communication or 3D touch, refers to the use of touch to enable humans to control and interact with computers through motion, force, and even vibration. Those of us who use touch screens to scroll through the photos on our phones or our social media feeds are already using haptics in our everyday lives. But a team in the mechanical engineering department at Texas A&M University, led by Professor Cynthia Hipwell, is working to revolutionize the virtual touch experience by developing technology that will allow a computer screen to "mimic the feeling of physical objects".[637] As an example of what the future of haptics may hold, imagine you're sitting on your sofa, doing a little shopping for some new pieces to add to your winter wardrobe. Or, maybe you're not really shopping so much as browsing a few websites—the modern-day equivalent of what used to be called window shopping—and you come upon what looks like a truly viable candidate for your closet: a chocolate brown, cable knit turtleneck. You like the way it looks on the model in the site's photo—its chunky fit and generous collar seems as if it would be cozy to wear in the cold months ahead. You

---

[637] "Better Touch Screens Could Let You Feel Stuff Before You Buy It," Texas A&M University (October 21, 2021), https://www.futurity.org/touch-screens-mimic-feeling-objects-2645982-2/.

like what you read in the product description—the fiber content, and the agreeable price point. You just wish you could feel the sweater so you could decide if it's as soft as it looks like it might be. In the future, if the team at Texas A&M is successful, you'll be able to simply put your fingers on the computer screen and "feel" the fibers for yourself, before you commit to the purchase—and we will have uncoupled the convenience of online shopping from the hassle of returning goods that disappoint once they're actually in your hands.

Virtual reality may, however, lead us into darker territory. A metaverse—a term coined by Neal Stephenson in his novel *Snow Crash*—is, in the simplest understanding, a space where humans can interact in a shared virtual reality. Gamers are already familiar with the concept and know it as being Massively Multiplayer Online (MMO). They inhabit their shared virtual space somewhere on the World Wide Web as their chosen avatar and talk to each other, play games, and generally hang out. The concept of *the* Metaverse—capital "M'—is also of a shared virtual reality, but one in colored, textured 3D that *all* humans, in the form of avatars, can inhabit; the most vivid way to think of it is as a shared alternate reality, or as what William Gibson named it in his seminal 1984 novel, *Neuromancer*, a *consensual hallucination*. It's an endlessly connected virtual universe, a reality created neither by physical science nor the divine, but by human dreams and desires, where ids are free to roam. It's where we'll meet to go to concerts and the theatre, to tour museums and cheer at sporting events and ride the rides at amusement parks. We'll play games together and talk to each other and generally hang out. We'll work there, go to school there, play there, and worship there.

Or, anyway, our avatars will do all of those things while we sit at our desks or on our sofas, strapped into virtual reality goggles and gloves.

And Mark Zuckerberg has announced he's building it.

In October, 2021, Facebook began rebranding itself as Meta, to reflect the new focus of the company's endeavors, and Zuckerberg promised the $10 billion he's investing in establishing the Metaverse would create a world "as detailed and convincing as this one" where "you're going to be able to do almost anything you can imagine."[638]

Now, some people think the Facebook rebrand is being undertaken to "serve to further separate the futuristic work Zuckerberg is focused on from the intense scrutiny Facebook is currently under for the way its social platform operates today."[639] In the most cynical interpretation, some think that the distance Zuckerberg is creating from Facebook is so that, as Facebook continues to encounter problems, Zuckerberg can simply say, "Call the CEO of Facebook, I'm Meta." Others think the Facebook brand is too damaged to be redeemed by a simple name change, even given the shift of focus away from being a social media platform and toward creating a new virtual world.[640] Still others believe he might succeed with the Metaverse venture, but—given that the same man who has been calling all the shots for Facebook will now be calling all the shots for Meta—the new Meta environment will be as toxic as the old Facebook one.[641] My own thoughts on the subject are that we are thinking about the wrong things.

There is something bigger at stake within the prospect of a Zuckerberg Metaverse—something grander than whether his new business venture will be a success or a failure, or even how we might feel about either outcome depending on whether we like Mark Zuckerberg as a personality or not. What we need to be considering is what will a Metaverse mean for us as human beings? What does living, working and playing, in a virtual reality universe do to the human condition? What will it mean

---

[638] James D. Walsh, "Why Facebook's Metaverse Is Dead on Arrival," *New York Magazine* (November 8, 2021), https://nymag.com/intelligencer/2021/11/why-facebooks-metaverse-is-dead-on-arrival.html.

[639] Alex Heath, "Facebook is planning to rebrand the company with a new name," *The Verge* (October 19, 2021), https://www.theverge.com/2021/10/19/22735612/facebook-change-company-name-metaverse.

[640] Eric Lutz, "Mark Zuckerberg Can't Wash Away Facebook's Toxic Image With a Rebrand," *Vanity Fair* (October 20, 2021), https://bit.ly/3wJG3Th.

[641] Emily Jane Fox and Joe Hagan, "Will The Metaverse Be Filled With Nazis?", *Vanity Fair* (November 5, 2021), https://www.vanityfair.com/news/2021/11/will-the-metaverse-be-filled-with-nazis.

for our relationships and our social interactions if we go on a date as an avatar, *with* an avatar, or watch our child's avatar play baseball on a virtual diamond? How will, for example, taking a hike on a virtual mountain trail instead of climbing an actual mountain impact on human health and evolution? When we are living our lives sitting in chairs or on sofas, heads gripped by goggles and hands shrouded in gloves, what becomes of those things we no longer actually see, or actually touch? What will be the consequences for our society? Our culture? Our democracy? The Metaverse, if, or *when*, it comes, will be enormous and we can no longer afford to continue asking the small questions.

o o o

Keep in mind, however, that this is just a taste of what might come to be in the years and decades ahead. Ray Kurzweil, famed futurist and inventor—who has an uncanny record of being right—has said that "people ignore the exponential nature of technological change…. [I]f you leave out the exponential, thirty changes get you to thirty…[but] thirty doublings in an exponential gets you to a billion."[642] The explosive nature of tech innovation—one advancement leading to the next, and then the next, each building on what has come before—will only increase as more innovations take us to more unheard-of places in less time than ever before in human history. We have, in our lifetimes, the promise of new wonders that will skyrocket and then fizzle to Earth, like Netscape and Myspace, as well as those with the staying power to change our lives for generations to come.

And, make no mistake about it, people are eager to sort out the fleeting from that with staying power—and the desirable from the not—among those innovations that seem to be making themselves a more permanent presence in our everyday reality. In June 2021, Data for Progress, a think tank for the future of progressivism, polled likely voters

---

[642] Ray Kurweil, "The Future of Intelligence, Artificial and Natural," *YouTube* (November 3, 2019), https://www.youtube.com/watch?v=Kd17c5m4kdM.

to find out their views around tech issues.[643] What they found was, in many respects, unsurprising. For example, protecting children from the exploitation of their personal data by online predators was at the top of the list for 88 percent of these potential voters; 86 percent supported the creation of new rules and standards for major technology companies in order to protect children's physical and mental well-being; and 89 percent supported the idea that additional measures should be imposed on those tech companies to stop the spread of child pornography. When asked how they felt about data privacy, these voters once again spoke clearly: 88 percent were concerned about foreign governments breaching data infrastructure in the United States; 85 percent were concerned about the spread of disinformation on social media; and 85 percent were worried about the collection and sale of their personal information. Moreover, these voters were, by large margins, disturbed by security missteps by, and other allegations against, social media platforms: 85 percent were disturbed that Facebook has been cited as the largest source of child pornography reports; 83 percent were concerned that Facebook has been found to engage in "friendly fraud," such as tricking children into making in-app purchases on their parents' credit cards; and 81 percent were concerned that Instagram, which is owned by Facebook, has exhibited a high incidence of cyberbullying on its platform.

The statistics I'm citing here are only the highlights of a more comprehensive report; I highly recommend you go to the link in the footnotes and read the whole report for yourself. Meantime, the takeaway is that a vast majority of American voters have had it with letting Big Tech play fast and loose, and they are ready to take action to rein them in. They're done allowing Big Tech to define the technological landscape in which we live, and they want the needs and desires of the people to be factored into the decisions that the Five are, so far, making unilaterally on behalf of all of us.

---

[643] Brian Burton and Ethan Winter, "Voters Want to Take on Big Tech Companies," *Data for Progress* (June 2021), https://www.documentcloud.org/documents/20950023-voters-want-to-take-on-tech-companies-2.

o o o

In 1896, a woman named Bridget Driscoll was crossing Dolphin Terrace in the grounds of the Crystal Palace in London when she was struck by a horseless carriage and killed. She was the first person to ever be killed by a motor vehicle in all the U.K., and the death was so unusual it triggered a full-blown inquest. Though witnesses described the car that hit Bridget as traveling at "a reckless pace, in fact, like a fire engine," the actual top speed of the car was eight miles per hour. A jury returned a verdict of "accidental death" in the ensuing case brought against the Anglo-French Motor Carriage Co., which was engaged in a demonstration of its cars when the fatal accident occurred. As the case was closed, the coroner remarked that he hoped "such a thing would never happen again."[644] Given that the National Safety Council puts the deaths from motor vehicle accidents in the United States for the year 2020 alone at 42,060,[645] we can safely say those folks long ago, in 1896, did not have any real understanding of how the new technology of horseless carriages was poised to change their lives. They did not allow themselves to thoroughly imagine how motorized vehicles could change their lives for either better or for worse.

Generally, we aren't engaging our imaginations today any more efficaciously than the folks back then, although there are signs of hope. For example, in the summer of 2021, Snapchat discontinued a feature they'd introduced in 2013 known as the "speed filter," which allowed users to "capture how fast they were moving and share it with friends." Safety advocates had been lobbying the company for years to remove

---

[644] Fredrick Kunkle, "Fatal crash with self-driving car was a first—like Bridget Driscoll's was 121 years ago with one of the first cars," *Washington Post* (March 22, 2018), https://www.washingtonpost.com/news/tripping/wp/2018/03/22/fatal-crash-with-self-driving-car-was-a-first-like-bridget-driscolls-was-121-years-ago-with-one-of-the-first-cars/.

[645] National Safety Council, "Motor Vehicle Deaths in 2020 Estimated to be Highest in 13 Years, Despite Dramatic Drops in Miles Driven" (March 4, 2021), https://www.nsc.org/newsroom/motor-vehicle-deaths-2020-estimated-to-be-highest.

the feature, citing crashes that had occurred while "drivers were moving at excessive speeds, allegedly to score bragging rights on the app," and families of those who had been killed or injured as a result of the crashes had brought many lawsuits against the company. When Snapchat finally removed the app, it gave disinterest as the reason: the feature "is barely used by snapchatters. And in light of that, we are removing it altogether," a spokesperson for the company said.[646] It may have taken seven years, and the company may have claimed the reason was disinterest, but the bottom line is that the speed filter was removed. There were unacceptable consequences to a particular piece of technology, and so the technology was rejected.

Good start.

Our continuing task is to imagine the consequences of the various technologies we are allowing into our lives, and to decide, both individually and collectively, if we will embrace what they are offering or if we will decline. It's true that most of the decisions we will need to make in relation to tech are not going to be as clear-cut as the decision to reject Snapchat's speed filter. They're going to require much more imagination in order to see their broader implications—what actually might await us as we adopt them and begin to integrate them into our everyday lives.

As I was writing this book, I read a story in the *Washington Post* about the demise of the circus.[647] It was an elegy for the passing of what was once an American institution—the spectacle of strongmen and sequined ladies and exotic animals riding into lonely, rural towns across the land and, for a day or two, offering wonder. "The cry 'the circus is coming to town' once signaled a fourth major holiday, equivalent with Thanksgiving, Christmas and the Fourth of July. Shops, public offices

---

[646] Bobby Allyn, "Snapchat Ends 'Speed Filter' That Critics Say Encouraged Reckless Driving," *NPR* (June 17, 2021), https://www.npr.org/2021/06/17/1007385955/snapchat-ends-speed-filter-that-critics-say-encouraged-reckless-driving.

[647] Les Standiford, "The disappearance of the circus from American life leaves us lonelier," *Washington Post* (June 20, 2021), https://www.washingtonpost.com/outlook/2021/06/20/disappearance-circus-american-life-leaves-us-lonelier/.

and schools closed, and an entire populace assembled to witness the parade of bands, clowns, exotic animals and bejeweled performers marching from the rail yards to the circus grounds, paced by aromatic elephants and shrieking calliope music all the way."[648]

It was also an invitation to meditate on those things that are passing away, or that might pass away, in our own time. If the arrival, and eventual ubiquity, of the technological wonder of motor cars made it easy for people in rural America to travel for entertainment, in the process making the circus and all of its glories passé, what is the arrival of so much new technology in our own era going to mean for us? What was—what *is*—going to pass out of our hands and into history? What good parts will we lose along with the bad? No one mourns that elephants are no longer held captive, in chains, on the back lots of dusty circus grounds; but we miss the jugglers, the fire-eaters, the clowns, the magicians—and the magic.

We have come to ask, more frequently as more and more of us disappear behind the screens of our computers and phones for longer and longer stretches of time, how technology is going to impact our ability to be social without the buffer of social media platforms. Now that it has allowed us to survive and even thrive through fifteen months of pandemic isolation, how will we break those bonds and emerge again into our larger communities? Will we go back to our schools and offices and once again gather around the proverbial water cooler, or will we remain in our homes, working and learning and socializing from home offices and kitchen tables, not wearing pants unless we're on a holographic call? Are we eager to return to "normal" life, as it was in the before times, or do we leave behind our temporarily tech-dominated community interactions with some anxiety? "A report by the American Psychological Association, published in March, 2021, found that almost half of Americans surveyed felt 'uneasy about adjusting to in-person interaction' after the pandemic."[649] Did the pandemic, then, only hasten an inevitable and full-blown

---

[648] Ibid.

[649] Anna Russell, "The Age of Reopening Anxiety," *The New Yorker* (June 3, 2021), https://www.newyorker.com/culture/dept-of-returns/the-age-of-reopening-anxiety.

dependence on communities facilitated by tech? After all, even prior to the onset of the pandemic, movie attendance was waning—down by 28 percent in 2019[650]; and American attendance at worship services hit an all-time low in March of 2021: "For the first time in eighty years, Gallup has found that less than half of U.S. adults belong to a church, synagogue or mosque."[651] Even before the virus struck, had we already been evolving out of our need for in-person, human-to-human contact? Is that even something a human *can* evolve from?

But complaining about human beings hiding themselves behind their cell phone screens to the detriment of the art of face-to-face conversation has, by now, become an old trope, even if it is wrapped within the relatively new context of a global pandemic. What are the broader ramifications of *emerging* tech? What does it do, for example, to our sense of home if we can travel from New York to Paris and back again in an afternoon on zirconium carbide-coated airplanes? Where is home, if home can be a different continent from one hour to the next? Is our community nowhere if it is also everywhere? Does cultural identity become a moveable feast in that case, or does it become lost in translation?

As science and technology continue to revolutionize our lives—often from year to year, and sometimes even from month to month—what will we retain? Will lining up around the block to score tickets to the latest blockbuster go the way of, well, Blockbuster?[652] Will going to the movies become a memory we hold, as our grandparents held the memory of the circus spectacle we'd never experience? What traditions will we decide we want to keep? Will we still enjoy the Macy's Thanksgiving Day Parade,

---

[650] Megan Cerullo, "U.S. Box Office Down 28 Percent This Year, But Not Because of Streaming," *CBS* (February 25, 2019), https://www.cbsnews.com/news/movie-box-office-down-28-percent-this-year-but-not-because-of-streaming/.

[651] Hannah Frishberg, "American church attendance hits historic low, says Gallup survey," *New York Post* (March 30, 2021), https://nypost.com/2021/03/30/american-church-attendance-hits-historic-low-survey/.

[652] Elizabeth Aubrey, "The World's Last Remaining Blockbuster Store Still Open Despite Coronavirus Pandemic," *NME Magazine* (May 14, 2020), https://www.nme.com/news/film/the-worlds-last-remaining-blockbuster-store-still-open-despite-coronavirus-pandemic-2668617.

even when it's a wholly holographic extravaganza? Will the Super Bowl be as exciting if we send AI-equipped robots out onto the playing field? Will group Zoom calls suffice in place of the annual family reunion weekend at the lake? What new myths will grow, out of necessity, from our new reality? How will we frame what we have gained, and what we have lost, in the next ten years? Twenty? In the next century?

Right now, in this moment, we have the power to decide how we will live with the wonders of the future. As I write, in 2021, privacy is—and is becoming ever more—a luxury item. If you are tech-savvy enough, and if you can afford to pay for premium privacy protection services such as an annual VPN subscription, you're going to be ahead of the game in retaining ownership of your personal data. For the vast majority of us, however, Big Tech is infiltrating our lives, shaping how we use its products, and exploiting our personal information for its own gain.

But privacy doesn't have to be a luxury good. Not if we speak up and start telling Big Tech how we will allow it to be part of our lives—which services and innovations we will accept as beneficial and will use, and which we will reject. We can stop their taking from us whatever it is they find useful, and, instead, make our own decisions about what it is we are willing to give. We *can* reclaim our digital dignity. We *can* choose to create a well-regulated, human-centered technological present that can and will make the wonders of the future positive ones for individuals, as well as for our society. The information and the guidance to make it so—well, you're holding that in your hands.

After all, it's your data. Shouldn't it be your billions, too?

Author's Note

"Received wisdom" refers, simply, to the way in which things are normally done—to custom based on "common knowledge". For example, many of us still *feed a cold and starve a fever* though whether or not the old wives who originated this advice were on target is debatable.

Dutch scientists conducted a small study to test the veracity of this conventional wisdom and found that gamma interferon, "a substance important in triggering immune responses against infection, particularly by viruses, increased by an average of 450%"[653] in study participants who ate before their blood was tested. Therefore, as colds are caused by viruses, continuing to eat when you're sick with one can help you to heal.

On the other hand, fasting—or, as the old wives had it, starving yourself—"appeared to increase levels of another immune system signaling chemical called interleukin-4 on average fourfold".[654] Interleukin-4 "plays a key role in fighting bacterial infection"[655] and, as many fevers are caused by bacterial infection, fasting can help with the healing process.

Done and dusted then, right? The old wives were right.

Well, not so fast. Many fevers are caused by viruses too—and, anyway, there is some debate in the linguistic community that the whole

---

[653] Claudia Hammond, "Feed a cold, starve a fever?", *BBC Future* (December 2, 2013), https://www.bbc.com/future/article/20131203-feed-a-cold-starve-a-fever.

[654] Ibid.

[655] Ibid.

proverb is a mistranslation and the original meaning was that "feeding a cold would 'stave off' a fever".[656]

So much for this bit of conventional advice.

o o o

Possibly the only received wisdom relating to Big Tech is that it is a fast-paced industry that offers us users rapid and often stunning innovation on a regular—sometime monthly, or even weekly—basis. The vast majority of the world's population is now connected, thanks to the advances of Big Tech and, over the last twenty years or so, have come to appreciate the convenience and pleasures technology offers to us.

Many of us have also come to accept that there are major downsides intermingled with all of this convenience and fun—but that a full understanding of all of them is beyond the typical user. That the only people who should try to understand the intricacies and finer points of all this massive innovation are actual techies—engineers and developers and tech ethicists and such—who have the practical and educational backgrounds to more easily comprehend these advances, the technology that enables them, and what each bit of technological progress will mean for individual users as well as the human race as a whole.

The rest of us are just supposed to integrate the advances into our lives—or not—and adapt. Big Tech is just too big and powerful. Trying to regulate what the tech industry does—or what it has planned for us—would be like trying to put a genie back in a bottle. Battling the kraken. Going to a gun fight armed with only a knife.

Presumably you picked up this book because you believed this particular received wisdom was...*debatable*.

The underlying discomfort we feel about Big Tech is well-founded. The changes it has brought, and is inevitably bringing, into our lives are going to have huge implications for the society, our culture, our government and governments around the world. While, however, we may never have, an individuals, a broad comprehension of the intricacies involved in

---

[656] Ibid.

how tech actually works, we can see the changes around us, sense what is to come, and know intuitively that it has to be us, collectively, who decide how tech is going to fit within our human world.

Because the idea of humans fitting into a world dominated by tech is unacceptable.

Now that you've read this book, I hope I've dispelled any lingering doubt that the received wisdom about Big Tech is, at its core, open for interpretation. We *can* take the baton and set the tempo for tech, but only if we, as consumers, do two simple things. 1) Live our online lives with discernment; and, 2) remain well informed about the constantly changing tech landscape around us so that we can accomplish number one with intelligence and dignity.

o o o

As we go to press with this book, it is becoming harder and harder to keep up with breaking news about tech. The wonders and the threats that come along with the creation of the Metaverse, the emergence of extreme social media platforms, the increasing prevalence of AI that is smarter than any human will ever be and too powerful to fight against, the continuing bumbling federal response in the US and the, somewhat, more informed one of the EU and other, smaller countries—it can all feel overwhelming.

So, I want you to keep this in mind as you follow the developments and innovations to come: what I've said in this book remains your base line. Judge each development, each innovation, each new digital revolution and each new piece of Big Tech-related legislation based on how it impacts upon your online privacy and your family's online safety. Yes, allow yourself to be rightfully dazzled at the brilliance of what engineers are more and more able to offer to us, but keep a grain of salt handy. Invention may belong to the creators within Big Tech, but how, and *if*, we will accept the next and the next best thing? Well, that is, at the end of the day, wholly up to *us*.

Let that become your new received wisdom.

Acknowledgments

Like it or not, we are living within an increasingly technological landscape.

The hard truth is that no matter how nostalgic we may feel for times we have—or, perhaps, have *not*—experienced in the past, the future is a technological one. We cannot go back to a time before we communicated through email and Zoom, or before we shopped so regularly—even habitually—online, or before we even knew what social media was, no matter that we might, from time to time, very much like to do just that.

According to a poll conducted by John Della Volpe, author of *Fight: How Gen Z is Channeling Their Fear and Passion to Save America* and the director of polling at the Harvard Institute of Politics, 68 percent of Americans—nearly two-thirds of the country—agree with the statement that "life was better before social media."[657] What is truly heartbreaking about his findings, however, is that 53 percent of Gen Z members agree that life was better before social media too[658]—and the members of Gen Z, that is, people born between 1997 and 2012, have never known life without social media.

Think about that: over half the members of the generation that is now coming of age in America believe that life was better within the social conventions that existed *before they were born.*

---

[657] "Nearly Two-Thirds of Americans Agree Life Was Better Before Social Media," MSNBC, October 12, 2021, https://www.youtube.com/watch?v=rn3zt-xMCBU&list=WL&index=5.

[658] Ibid.

But the ecosystem in which we function as a community has shifted—seismically—and that means we must shift our perspective too. Our challenge isn't to fear the unknown or to bemoan the change, but to adapt to the new landscape.

The trick—the responsibility—is to adapt in ways that serve us as human beings, and as a society, and it will be quite a trick to pull off: Big Tech, the very people who are providing us with the hardware and the software that make our progressively more conveniently technological lives possible, are public corporations. Their charter is to make money for themselves and their investors, not to provide goods and services that best serve our human community.

So, it is up to us to figure out how to live dignified online lives—and how we will regulate those corporations so this goal is possible. In writing this book, I hoped to provide that perspective. To outline what lies ahead of us, and to create a guide that simplifies the often-overwhelming tech world for those of us who share this goal. I am grateful to the many people who gave help, encouragement, and provided inspiration along the way.

Thank you to Patti Bellinger, Mike Emery, Paul Fedorko, Antonia Felix, Scott Galloway, Melinda French Gates, James Gregorio, Daisy Hoffman, Katie Hoffman, Michael Hoffman, my Harvard Gender Equity co-chair Doris Honold, Reed Hundt, George Igel, Alice Martell, Nicco Mele, Kate Monahan, Bryan Panzano and the Harvard 2020 Advanced Leadership fellows, and the whole Harvard Advanced Leadership Team, Alexandra Penney, Bruce Rich, Meredith Rosenthal, Kara Swisher, Lynn Vannucci, Jim Waldo, Megan Wheeler, Tom Wheeler, and Anthony Ziccardi.